FOR WREN +
NIDHI +
FRIENDS,
HORACE

UNDERSTANDING DNA

DNA

The Molecule & How It Works

UNDERSTANDING DNA

DNA

The Molecule & How It Works

by

C.R. Calladine

Department of Engineering
University of Cambridge, Cambridge, UK

and

Horace R. Drew

Laboratory for Molecular Biology
CSIRO Division of Biomolecular Engineering, Sydney,
Australia

ACADEMIC PRESS
Harcourt Brace Jovanovich, Publishers
LONDON   SAN DIEGO   NEW YORK   BOSTON
SYDNEY   TORONTO   TOKYO

ACADEMIC PRESS LIMITED
24–28 Oval Road
London NW1 7DX

US edition published by
ACADEMIC PRESS INC.
San Diego, CA 92101

A catalogue record of this book is available from the British
Library

ISBN   0–12–155085–0
          0–12–155086–9 pbk

Typeset by Photo·graphics, Honiton, Devon
Printed in Great Britain at The University Printing House,
Cambridge

Contents

Preface

We also now appreciate that molecular biology is not a trivial aspect of biological systems. It is at the heart of the matter. Almost all aspects of life are engineered at the molecular level, and without understanding molecules we can only have a very sketchy understanding of life itself. All approaches at a higher level are suspect until confirmed at the molecular level.

Francis Crick, *What Mad Pursuit*, 1988

This is a book about DNA, the most central substance in the workings of all life on Earth. It is a book about the way in which DNA works at a molecular level. We have used the title *Understanding DNA* ... because our subject has now reached the stage where many aspects of it are well enough understood for us to be able to give a clear and uncluttered presentation of the main ideas. But we shall not disguise the fact that there is still a great deal which is not known or understood.

The book can be read at two different levels. First, it can be taken as an easy-to-read textbook for students of chemistry and biology at university. Second, it may be read by ordinary people who have no prior knowledge of biochemistry, but who want to understand something of the fundamental processes of life. The sort of people we have in mind are those who have learned something about DNA from popular magazines, newspapers, and TV programs. They know, for example, that DNA contains the "genes" of classical genetics—those units of inheritance which pass on characteristics such as red hair or a long nose from parent to child, or even crippling diseases such as sickle-cell anemia or thalassemia. They probably also know that DNA is a long molecule, like a computer tape—the tape which tells our bodies how to grow and how to digest food and (perhaps) how to behave. And they may even know, if they are into quiz games and the like, that the initials "DNA" stand for "Deoxyribo-Nucleic Acid," a certain kind of acid found in the cell nucleus, which was first identified over 100 years ago. People like this, who are curious to know more, will be able to learn a lot from this book about how DNA performs its tasks in our bodies at a molecular level.

The book comprises eight chapters. Chapter 1 is a general introduction to molecular biology: it is aimed at the non-specialist reader, and so it may be passed over by a student who already knows some biology. Chapters 2, 3, and 4 give some lessons about various aspects of the molecular structure of DNA, such as why it is helical, and how it can bend; this is basic material, which is nevertheless not yet available in other textbooks. Chapters 5 and 6 discuss the three-dimensional structure of DNA at a higher level. These chapters include some mathematics and geometry that may be unfamiliar to non-specialists and biology students; but we take care to present the ideas by means of clear diagrams wherever possible. Chapter 7 gives an overview of the organization of chromosomes, which are large particles that contain both protein and DNA: there the DNA wraps about the protein into several different levels of structure. Lastly, Chapter 8 explains the various experimental techniques which scientists use to study DNA. We end with a Postscript on what we have left out, followed by two Appendices on matters too detailed for the main part of the book. We have also provided some further reading for the student and a few exercises at the end of each chapter.

Understanding DNA: the molecule and how it works should be suitable as a small text to accompany the very large, general textbooks which are now used widely in university biochemistry courses; or else it may be employed as a main text for a course specializing in DNA structure, provided the students are willing to pursue more detailed readings in the scientific literature, as suggested. Or, of course, it may just be read as a book.

Many friends and colleagues have helped us greatly in various ways in the preparation of this book. We are grateful to Nick Cozzarelli, Malcolm Ferguson-Smith, John Finch, Ron Hill, Chris Hunter, Louise Lockley, Ben Luisi, Maxine McCall, Margaret Mott, Dinshaw Patel, Tim Richmond, David Tremethick, and Sue Whytock for providing photographs and diagrams which we have used; to Dick Dickerson for giving us data on X-ray structures; to Tina Collis, Aaron Klug, Daniela Rhodes, Jean Thomas, and Michael Waring for commenting freely on various drafts of the manuscript; and to Julian Wells and Carole Gannon for encouraging us in the first place to write a book on DNA. Susan King has been a constant source of editorial advice and encouragement. The work of Marica López Díaz in typing the

manuscript and of Dennis Halls in making the diagrams has
been beyond praise. Lastly we thank our respective wives, Mary
and Maxine, for their help of many kinds over the years; and
we dedicate the book to them with gratitude.

C.R.C.   H.R.D.
Cambridge   Sydney

CHAPTER 1

An Introduction to Molecular Biology for Non-Scientists

One day we were having lunch together at a Cambridge College. We got into a conversation with one of our neighbors at the table, who was a senior historian. After a while he asked us what we did, and we explained that we were scientists, working with the very tiny molecules of biology. Then he said, "I don't see how you do it."

"Do what?"

"Work all the time with things that you can't see."

You see, even people of great intelligence and learning, who spend their lives gathering evidence and pondering it deeply, nevertheless think in ways very different from those of modern biologists; it is hard for them to imagine what atoms and molecules look like. How hard will it be, then, for the beginning student to do the same?

For this reason it is necessary for us to start this book by explaining carefully about the *sizes* of things. A single DNA molecule is too small to be seen by eye. But if you have a big clump of lots of DNA molecules together, then the substance becomes visible, and appears as a clean, white, stringy, and viscous mass, somewhat like molasses sugar. Yet, you *can* see single DNA molecules by using special equipment involving X-rays, or an electron microscope; and we shall show some pictures of these later in the chapter.

But first, in order to gain an intuitive feeling for the microscopic world, let us compare the size of DNA to the sizes of things in general, and especially to the sizes of other things that are too small to be seen by the naked eye. Figure 1.1 shows a scale of typical sizes. It is a logarithmic scale, and each division represents a factor of 10. The scale covers 10 orders of magnitude, from 1 m down to 0.000 000 000 1 or 10^{-10} m. Near the top we

LENGTH (m)			$10^6 \times$ LENGTH
		Total length of DNA in a human cell	
1	1 m	Length of a human being	1000 km
10^{-1}			100 km
		Length of DNA from a typical human chromosome	
10^{-2}	1 cm		10 km
10^{-3}	1 mm	Length of DNA isolated from a bacterial chromosome	1 km
10^{-4}			100 m
		Length of a typical human chromosome in extended form	
10^{-5}		Diameter of a typical human cell	10 m
		Length of a typical human chromosome in compact form	
10^{-6}	1 μm		1 m
10^{-7}		Diameter of a solid ball containing human DNA	
10^{-8}			1 cm
		Diameter of a DNA thread	
10^{-9}	1 nm		1 mm
10^{-10}	1 Å	Diameter of an atom	

Figure 1.1 The relative lengths of things on a microscopic scale.

have the largest objects that we shall be thinking about: human beings are roughly 1 m long, as an order of magnitude. At the bottom of the scale are the smallest objects that we shall be concerned with: atoms, which are typically of diameter 10^{-10} m (or 1 Ångstrom unit, Å). Exactly halfway between these two extremes, on the present kind of scale, we have the diameter of a typical human cell at about 10^{-5} m or 10 μm.

Some things are larger than a cell on our scale, while others are smaller. It is perhaps surprising that the length of DNA isolated in pure, thread-like form from a single human chromosome[1] (3 cm, or 3×10^{-2} m) or from a single bacterium (1 mm, or 10^{-3} m) can be so much longer than the cell from which it came (10^{-5} m). But this illustrates a very important point: the DNA is compacted in length by a factor of as much as 10 000 when it is embedded in a living cell. DNA is a very narrow thread of diameter just 2×10^{-9} m, and although the DNA from a single human cell has a total length of 2 m, it could conceivably be compacted into a tight ball, like a ball of string, of diameter 10^{-7} m. But Nature chooses to pack the DNA into a form somewhat less dense than this, at a length of about 3×10^{-6} m for a single, compact chromosome. After all, if DNA were packed too tightly into a cell, then the information along its length, known as "genes"[1] (to be discussed below), would probably become inaccessible.

Our bodies are made from billions of individual cells, and DNA is the control-center of each and every cell. This DNA is something like the computer tape that stores many programs for a large computer to run. It is present in each cell in the form of a number of chromosomes. Chromosomes are arranged in pairs (see Fig. 1.2), and the two members of a pair are nearly-identical or "homologous"[1] copies of one another, just like the back-up disks on a computer: one copy saves the functional program if the other copy becomes defective.

The mass of DNA is surrounded in most cells by a strong membrane with tiny, selective holes, that allow some things to go in and out, but keep others either inside or outside. Important chemical molecules go in and out of these holes, like memos from the main office of a factory to its workshops; and indeed the individual cell is in many ways like an entire factory, on a very tiny scale. The space in the cell which is not occupied by DNA and the various sorts of machinery is filled with water.

On the right of Fig. 1.1 is a second scale of lengths which would apply if we were to enlarge every linear dimension by a factor of 1 000 000 or 10^6. It can be useful sometimes to make an imaginary enlargement of this sort. When we do so, the relative sizes and proportions of objects remain the same, of course. Note that the length of DNA from a typical chromosome on this expanded scale is about 30 km, while its diameter is just 2 mm. Very few objects in the physical world are so long and so narrow.

Now that we have gained a general idea of the relative sizes of things, and of the importance of DNA in the control of a

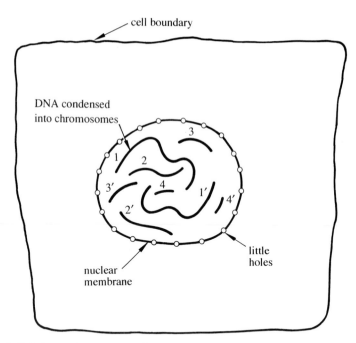

Figure 1.2 Schematic picture of a typical cell from higher organisms. The chromosomes, which contain a mixture of protein and DNA, come in homologous pairs. This picture shows a hypothetical cell with four such pairs, that can be distinguished by their sizes. They are labeled 1,1';2,2'; etc.

living cell, let us see what DNA and chromosomes really look like.

Human chromosomes become compact and squat when cells are about to divide, and they can then be seen easily by use of the light microscope. The human chromosomes shown in Fig. 1.3 have been sorted and arranged by size into pairs. Each chromosome, as in Fig. 1.2, has duplicated itself in preparation for cell division, and has then reduced its length 10-fold, so that the duplicate copies can separate from one another without tangling as the cell divides. One half of each X-shaped duplicate chromosome will go to each new cell.

Many more chromosomes are shown here than in the simple, schematic drawing of Fig. 1.2. Human cells contain 46 chromosomes in 22 homologous pairs (numbered 1–22) plus the non-homologous X and Y chromosomes that determine sex. All animals and plants have chromosomes that look like these, but in different numbers: for example, a fruit fly has 8 chromosomes, in 3 homologous pairs plus X and Y. Spreads of chromosomes such as those shown in Fig. 1.3 are very useful for medical

Figure 1.3 Photographs of human chromosomes in duplicate, as isolated just before cell division (at metaphase) and then sorted by length into pairs. Each number identifies the two chromosomes of a homologous pair. X and Y are non-homologous chromosomes that determine a person's sex as female (XX) or male (XY). The two duplicate copies of any individual chromosome form an "X" shape because they have not yet separated entirely. Scale: chromosomes 3 are approximately 10^{-5} m or 10 μm long. Courtesy of Malcolm Ferguson-Smith.

purposes; for example, to check the health of an unborn child while still in the womb.

At other stages in the life of a cell, far from cell division, the chromosomes are generally more extended and less condensed, and so they cannot be seen by use of a light microscope. That is why we had to draw the single chromosomes of Fig. 1.2 in schematic form, because no true pictures of such chromosomes exist. Nevertheless, in a few particular tissues of the fruit fly, these single-copy, extended chromosomes happen to duplicate themselves about 1000 times over, without becoming compact. They eventually contain 1000 identical DNA molecules laid side-by-side, in parallel register. These "polytene" chromosomes are just like the ones shown in Fig. 1.2, but they are much wider. Because of their greatly increased size, these monstrous, amplified fly chromosomes can be seen clearly by use of a light microscope, as shown in Fig. 1.4 at different magnifications.

In Fig. 1.4(a) the extended, DNA-containing chromosomes can be seen along with the tight membrane in which they are wrapped; the entire assembly is known as the "nucleus."[1] Here the chromosomes are like little worms or eels surrounded by water and trapped inside a bubble. When we break open the bubble and look more closely, as in (b), (c) and (d), we can see that each chromosome is divided along its length into many clear striations of dark and light, which are known as "bands" and "interbands." The bands are dense clumps of protein[1] plus DNA, whereas the interbands are sparse regions of low density. No one knows why the variations in dark and light are so regular, and indeed so reproducible from one fly to the next. Presumably, these divisions are marked off by certain patterns in the atoms along the DNA thread. But it makes good sense in biological terms to divide a chromosome along its length: after all, computer tapes store information in the same sort of way, as discrete files, each with a beginning and an end, along the length of a tape.

At a finer level, more is known. Each thread of the band and interband material in fruit-fly chromosomes (or indeed in human chromosomes) is composed of a well-defined mixture of protein and DNA. This protein is of a special kind that makes "spools", and the DNA wraps twice around each spool as shown in Fig. 1.5, into a series of double loops. The wrapping of DNA twice about each spool reduces its overall length by a factor of about 6.

When we remove the protein spools, we are left with a very long, string-like DNA molecule. The DNA from the longest individual human chromosome, if it were enlarged by a factor of 10^6, so that it became the width of ordinary kite string, would extend for about 100 km. Imagine sitting in a train traveling from Cambridge to London, or from Los Angeles to San Diego, and looking out of the window for the whole trip at a single DNA molecule and watching the genes go by!

The basic form of this immensely long DNA fiber is shown in Fig. 1.6. It consists of two strands which coil around each other to make a "double helix." The term "helix" is just another word for "screw" or "spiral." The sense of wrapping of these two strands is usually clockwise as you go forward, or right-handed—the same sense as an ordinary corkscrew.

If we uncoil the two strands, as shown in Fig. 1.7, then each

Figure 1.4 Polytene chromosomes from the salivary gland of a fruit fly. Length scales: 100 000 Å or 10 μm. Courtesy of Ron Hill and Margaret Mott.

protein spool   two loops

DNA

Figure 1.5 A DNA thread making two left-handed coils around each of a series of protein spools, as might be seen in the chromosomes of Figs 1.2, 1.3 and 1.4 if the magnification were higher. Each DNA–protein spool is about 100 Å (or 10^{-8} m) across.

Figure 1.6 DNA double helix, showing base-pairs. Now the DNA thread of Fig. 1.5 is shown in its true form, as a double spiral containing two chains of DNA. The diameter of the spiral or helix is about 20 Å.

strand may be seen to consist of a series of units called "nucleotides." These are linked to one another with a certain "directionality," known technically as "5-prime to 3-prime," in a head-to-tail sense that we shall explain below. The two strands run in opposite directions, as shown by the labels 5' and 3', and by the arrows.

Any nucleotide is made of about 20 atoms such as carbon, nitrogen, and oxygen. These atoms can again be grouped into smaller parts which are connected in a particular way. The three parts of a nucleotide are its sugar, phosphate, and base; in the diagram of Fig. 1.8 they are labeled S, P, and B, respectively. For present purposes we may draw the sugar as a five-sided

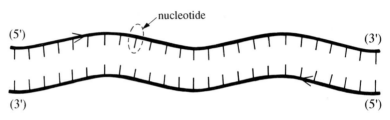

nucleotide

(5')   (3')

(3')   (5')

Figure 1.7 The two strands of DNA separated, showing a nucleotide. Each nucleotide is about 6 Å wide.

ring, because the atoms in a sugar join to form such a ring. We have taken the liberty of drawing the ring in the form of an arrow, in order to indicate its directionality; and for the same reason we have put the letters P, S, and B so that they read in this same direction from left to right. If we took out the central nucleotide in Fig. 1.8, rotated it through 180° about a vertical axis and tried to re-connect it, we would find that it would not link up.

We can draw the base as a rectangle because the atoms there join to form a flat, rigid shape. There are usually four different kinds of nucleotides in DNA, which share the same sugars and phosphates but have different bases attached to the chain. These four different bases form pairwise interactions that join the two strands of DNA together weakly (see Chapter 2).

The three parts of a nucleotide can also be studied in isolation, as collections of just a few atoms. For example, a phosphate contains one phosphorus and four oxygen atoms, as shown in Fig. 1.9. The atoms themselves are made of protons, neutrons, and electrons. They are connected together by the sharing of electrons; and the various kinds of chemical bond that connect atoms to each other are all aspects of this. We could go on to discuss atoms within the realm of subatomic physics, but there would be no point in this, because our knowledge of living things begins with chemistry (or biochemistry) and extends into biology. So we can stop our description of DNA with the relatively simple picture of atoms shown in Fig. 1.9, without losing anything.

If you have not studied biology before, you may be puzzled that we have put so much emphasis on the *cell*. Would it not be more sensible to start with the parts of the body, such as limbs, eyes, and lungs? The answer to this is that all of these

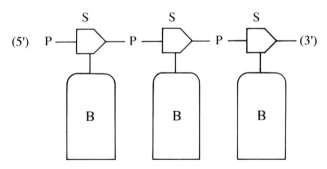

Figure 1.8 Three nucleotides, showing their sugar (S), phosphate (P), and base (B) components.

Figure 1.9 The phosphate group as part of a sugar–phosphate chain. Each atom is 1–2 Å wide; atom types are phosphorus (P) and oxygen (O).

various organs and tissues, etc. are built up from cells by essentially the same process.

You are probably familiar with the way in which a house is built. The various components—bricks, cement, tiles, timber planks, window-frames, etc.—are first delivered to the building site and then assembled in accordance with the architect's plan or "blueprint." This plan is a sheet of paper on which are drawings of the finished house taken from several points of view. There may well be only one copy of this plan, and it may be kept in the pocket of the foreman-builder and taken out and consulted from time to time, as required. Such a scheme of construction is typical for non-living things like houses, motor-cars, and gadgets.

In contrast, the scheme of construction for living things—whether they are plants or animals—is different from this in almost every way. Thus, construction of a human begins with a single cell—an egg from the mother which has been fertilized by a sperm from the father. This composite cell contains DNA from both the mother and the father; such DNA contains the complete genetic information for construction of a human being. Growth occurs by the process of cell division: each cell divides into two new cells, and these cells in turn divide, and so on. Just before any cell divides it duplicates all of its DNA, so that every new cell contains a complete set of DNA, which again contains all the genes of the organism. (We saw some pictures of duplicated chromosomes in Fig. 1.3.) Only a small fraction of all the genes present on this DNA are activated in any given type of cell. Thus, cells which develop into an eye use only the genes which program for the growth of eye-cells. How cells "know" which kind of organ they belong to is a large and poorly understood area of research, *developmental biology*, which we shall not go into here.

We mention all of this because it may seem, to a non-scientist, to be enormously wasteful for Nature to provide a complete set of DNA in every one of the billions of cells of every animal or plant: would it not be more straightforward and efficient just to have a single copy of the design information, just as in the construction of a house? However, a little thought indicates that the scheme for providing every cell with a complete set of DNA is, in fact, an extremely simple way of providing this necessary information, in all places where it is required—even though, of course, the scheme requires a vast amount of repetitive copying and duplicating of DNA. The machinery for duplicating DNA is accurate enough for the entire scheme to work well, and the double-helical structure of DNA is extremely convenient, as we shall see in Chapter 2, for the purposes of duplication.

We are now ready to clarify a further point: how does DNA carry the information necessary to run the activities of a cell as a factory, as its control-center or main office? The most basic way in which DNA runs the activities of a cell is to specify the composition and structure of protein molecules. Proteins come in a wide range of shapes and sizes, and play a wide range of roles in the life of a cell. Some proteins are strong and rigid, and form the building-components for muscles, tendons and finger nails. Other proteins, or "enzymes,"[1] catalyze a large number of chemical reactions such as digestion of food or the synthesis of hormones. Other proteins carry oxygen in the blood, while still others form the protein spools around which DNA wraps in a chromosome. Where do all these proteins come from? How does a cell know which proteins to make?

It turns out that the DNA runs a very definite "program" every time the cell needs to make a given protein molecule. The program tells the cell exactly what kind of protein to make, and approximately how much of it. This program is the well-known "gene" of classical genetics; and each DNA molecule contains many genes along its length, as shown in Fig. 1.10. Sitting in our train traveling from Cambridge to London, we would see thousands of them as we looked out of the window watching the DNA go by!

The proteins, although very different from one another in their physical and chemical properties, share a common scheme of construction: they are all long-chain molecules, consisting of a single, unbranched chain that is made from the end-to-end joining of many small units known as "amino acids." Proteins are made up from amino acids, just as DNA is made up from nucleotides. For example, the sweetener in Diet Coca-cola is a very tiny protein made by man from just two amino acids; while hemoglobin in the blood is a much larger protein, made protein 1   protein 2   protein 3

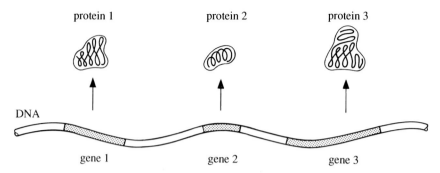

DNA

gene 1   gene 2   gene 3

Figure 1.10 Genes in the DNA code for proteins. The DNA thread shown here represents the double helix shown in Fig. 1.6, but here it is about 100 times longer. Both genes and proteins can vary in size, in direct proportion to one another. Each protein molecule is shown with a schematic chain-like structure.

from 100 amino acids. Altogether there are 20 different kinds of amino acid that make proteins; and this wide variety of building blocks allows for the construction of very many different proteins, with their enormous range of physical properties.

A program or gene in the DNA tells the cell in what order to assemble these amino acids and for what length of protein chain. The order of amino acids in a protein is set by the order of nucleotides in the corresponding piece of DNA. Because there are 20 possible amino acids, and only 4 possible nucleotides, the cell must use more than one nucleotide in the DNA to specify each amino acid in a protein. The universal rule is that *three* nucleotides specify *one* amino acid. The length of protein chain varies with the length of the DNA "program," and is typically from about 100 to 1000 amino acids.

The cellular machinery that enables a certain set of three nucleotides in the DNA to be associated with only one of the 20 possible amino acids is extremely complex. Likewise, the conversion of the magnetized stripes on a regular computer tape into printed characters on a page of computer output must be rather complicated. But we do not have to worry about this kind of machinery in order to explain the simple code or cipher by which DNA makes proteins. Usually, this cipher is presented in the form of a table called "The Genetic Code" (Table 1.1). This same code is used for specifying proteins in all living things, whether they are bacteria, plants or animals, with only a few minor exceptions.

The four different nucleotides in DNA are called adenine, guanine, cytosine and thymine, or simply A, G, C and T. The

Table 1.1 The Genetic Code

1st base	2nd base				3rd base
	T	C	A	G	
T	Phe	Ser	Tyr	Cys	T
	Phe	Ser	Tyr	Cys	C
	Leu	Ser	STOP	STOP	A
	Leu	Ser	STOP	Trp	G
C	Leu	Pro	His	Arg	T
	Leu	Pro	His	Arg	C
	Leu	Pro	Gln	Arg	A
	Leu	Pro	Gln	Arg	G
A	Ile	Thr	Asn	Ser	T
	Ile	Thr	Asn	Ser	C
	Ile	Thr	Lys	Arg	A
	Met	Thr	Lys	Arg	G
G	Val	Ala	Asp	Gly	T
	Val	Ala	Asp	Gly	C
	Val	Ala	Glu	Gly	A
	Val	Ala	Glu	Gly	G

Any series of three bases (or nucleotides) in the DNA prescribes for an amino acid in the protein chain, or gives a "stop transcribing" signal. The bases are always read from left to right. The chain usually starts with ATG or Methionine (Met). Abbreviations used: A, adenine; G, guanine; C, cytosine; T, thymine (or U, uracil in RNA). Ala, alanine; Arg, arginine; Asn, asparagine; Asp, aspartic acid; Cys, cysteine; Gln, glutamine; Glu, glutamic acid; Gly, glycine; His, histidine; Ile, isoleucine; Leu, leucine; Lys, lysine; Met, methionine; Phe, phenylalanine; Pro, proline; Ser, serine; Thr, threonine; Trp, tryptophan; Tyr, tyrosine, Val, valine.

20 possible amino acids in a protein have names such as "methionine," usually abbreviated to "met" or "M." As shown in Table 1.1, each possible set of three nucleotides in the DNA specifies one amino acid. For example, "TTT" specifies phenylalanine, while "AAA" calls for lysine, and "GCT" gives alanine. In each case the letters are read from left to right. There are $4 \times 4 \times 4 = 64$ combinations of DNA triplet, and often two of these triplets such as "AAG" and "AAA" specify the same amino acid. Thus, a series of nucleotides such as "TTTAAAAAGGCT" specifies a portion of protein with amino-acid sequence "Phe-Lys-Lys-Ala," as shown in Fig. 1.11. Certain triplets along the length of DNA also do the special work of telling the protein chain where to start and stop.

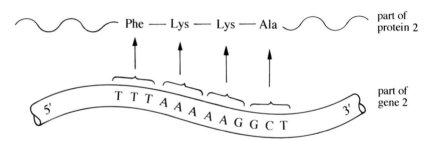

Figure 1.11 Part of the gene codes for part of the protein chain. Each set of three nucleotides (or bases) in the DNA specifies one amino acid, according to the scheme of Table 1.1.

Finally, the completed protein chain usually folds by itself into a precisely determined shape, which depends on the exact arrangement of the amino acids in its chain. A correct three-dimensional shape is essential to the physical or chemical activity of a protein.

No one should think that all DNA does is to make proteins like a piece of computer tape. In fact, only a small fraction of the long DNA molecule in a chromosome—about 1% in humans—contains programs to make specific proteins. The vast majority of DNA in our bodies does things that we do not presently understand. There is plenty of room here for people to make new discoveries.

We said above that we can safely ignore the complex machinery which the cell uses in order to make proteins. This is true for the most part, but we do need to know about a few features of the process. We described above how the DNA-containing chromosomes are surrounded by a membrane in the nucleus of every cell (except in bacteria, where there are no nuclear membranes). It has been discovered that proteins are made in the cellular space *outside* of this membrane. What happens is that the DNA of a particular gene first makes a copy of itself that can pass through the small holes in the membrane; then this copy goes off to the protein-making machinery, which is called a "ribosome." This process is shown schematically in Fig. 1.12. These copies of the DNA program are made from a slightly different kind of molecule called "RNA." At the level of the picture shown in Fig. 1.8, RNA has an extra oxygen atom on the sugar ring as compared with DNA. A second difference is that RNA lacks a carbon and three associated hydrogen atoms on the thymine (T) base, and so this base is renamed "uracil" (U). The RNA copy itself is called "messenger RNA," because it carries instructions on how to make a particular protein from the chromosome to the ribosome.

Interestingly enough, messenger RNA makes up only a small fraction of the total RNA in any cell, about 5% of the total; and there is also much more RNA than DNA altogether. A lot of RNA is copied from DNA but never used to make protein. These abundant RNA molecules are different from messenger RNA, and are mainly of two types: "transfer RNA" and "ribosomal RNA". Transfer RNA carries single amino acids from elsewhere in the cell to the ribosome; there the amino acids, while still bound to the transfer RNA, can pair up with specific triplets of nucleotides on the messenger-RNA chain, and so make a protein in accordance with the cipher of Table 1.1. In other words, transfer RNA performs the function of the vertical arrows shown in Fig. 1.11, once the DNA has been copied to messenger RNA. We explain more about this in Chapter 2. Ribosomal RNA makes up the bulk of the ribosome, in association with a few proteins; it helps to align a series of transfer RNA molecules along a chain of messenger RNA, so that a series of amino acids can be joined chemically to form a long protein. It seems remarkable that so many steps in the synthesis of protein involve an RNA intermediate. This observation has led many people to speculate that life on Earth began with RNA, rather than with DNA, as the substance of genes.

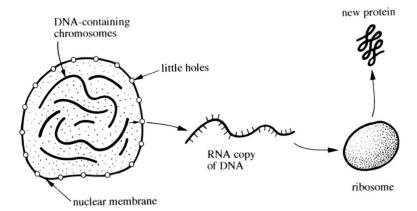

Figure 1.12 From DNA to RNA to protein. The DNA is copied into "messenger RNA," which then travels outside the nucleus to the ribosome, where it specifies the assembly of some particular protein, according to the series of nucleotides in its chain as in Table 1.1. For simplicity, we have not mentioned here that many genes in higher organisms contain substantial amounts of non-coding DNA, or "introns," that are cut out at the level of RNA before the RNA copy of DNA leaves the nucleus.

Now we have presented all of the essential background information that is required for you to understand how DNA works in biology. Other books could be written on RNA or protein, but here we focus exclusively on the role of DNA. We shall work from small to large, starting with the basic chemistry of DNA and ending with chromosomes. Despite much recent progress, the great revelations in biology will no doubt belong to the twenty-first century; we cannot pretend that our present knowledge is any more than a rough and incomplete foundation on which others can build.

Note

1. The linguistic or historical derivations of words marked thus are given in Appendix 1.

Further Reading

Crick, F.H.C. (1963) The recent excitement in the coding problem. *Progress in Nucleic Acid Research* **1**, 163–217. A good scientific review of how the Genetic Code was discovered.

Crick, F.H.C. (1988) *What Mad Pursuit*. Weidenfeld & Nicolson, London. An anecdotal history of the discovery of the structure of DNA and of the Genetic Code. Also, the source of the quotation used in the Preface.

Judson, H.F. (1979) *The Eighth Day of Creation*. Simon & Schuster, New York. A very thorough history of how molecular biology became a separate branch of science.

Watson, J.D., Hopkins, N.H., Roberts, J.W., Steitz J.A., and Weiner, A.M. (1987) *Molecular Biology of the Gene*, (4th edn). Benjamin/Cummings, Menlo Park, California. A good general reference volume, but not much on DNA structure.

Exercises

1.1 Every human cell, in the non-dividing state, contains a total of about 6×10^9 base-pairs of DNA. The diameter of a typical human cell is 10 μm $= 10^5$ Å.

   a   By treating the DNA as a cylinder of length 3.3 Å per base-pair, calculate the total length of DNA in any cell. Compare this to the diameter of the cell.

b By treating the DNA as a cylinder of radius 10 Å, calculate the total volume of DNA in any cell. Compare this to the total volume of the cell, if it is assumed to be spherical. (Volume of cylinder $= \pi r^2 l$, where r is the radius and l the length: volume of sphere $= \frac{4}{3}\pi R^3$, where R is the radius.)

c Consider a typical chromosome, which contains 1/46 of the total DNA of the cell, on average. Find the total length of the DNA in this chromosome, and then the diameter of a solid sphere into which that volume of DNA could be compacted, in principle. Compare the diameter of this compact sphere with the mean length of an actual, metaphase chromosome, given that the DNA length–compaction ratio for such a chromosome is about 10 000.

1.2a Use Table 1.1 to give the sequence of amino acids in a protein chain which is coded by the following base sequence:

GCCAAGCAACTCATTCAAGGT
123

Start reading at base 1.

b Now repeat the process by beginning to read at base 2; and then again by beginning to read at base 3. (Note that the amino acid sequence of a protein chain depends critically on which "reading frame" is used in the DNA.)

1.3a An extra base G is now inserted between bases C and T at positions 10 and 11 of the DNA sequence given in Exercise 1.2. Starting at base 1, read out the new sequence of amino acids.

b The sequence given in Exercise 1.2 is now altered instead by the deletion of C at position 10. Starting at base 1, read out the sequence of amino acids.

(These are both known as "frame-shift" mutations.)

1.4a Translate the following DNA base sequence into a sequence of amino acids for a protein molecule:

ACGCTATGTCACATGGTACCTAACGTAT

On this occasion, do not begin reading at position 1, but rather in accordance with the true, biologically used start-scheme described under Table 1.1, and do not read beyond a STOP triplet.

b Search the base sequence given below for potential STOP triplets. What amino acid sequence will be assembled, starting as in **a**?

GCTCATGGTCATTCGTAACAGTTAGGCCATGACCG

CHAPTER 2

Why a Helix?

It is crucially important, not only in biology but in all fields of science, to understand the inner workings of Nature as well as its external form. For example, consider the funny-looking object shown from two perspectives in Fig. 2.1 (see p. 21). It has five different colours, but what *is* it? How would you go about making such an object? A poor scientist would study the object superficially, and give a name to each particular feature such as the intersection of five edges; he would be deeply concerned with measuring the angles between edges on the surface of the object, in order to look for some sort of pattern. But a more perceptive scientist would stop and think: what is the internal structure of the thing?

Now look at the object again, but this time focus only on one color, say yellow: behold, the yellow parts form a cube in space, tilted somewhat on its side. Each of the other four colors also describes a cube. In fact, this toy is called "The Compound of Five Cubes."

Spotting the five distinct cubes—and it was helpful of the maker to use five different colors—is the first clue to understanding the internal structure of the object. This clue leads us to think about matters of geometrical symmetry, and if we were to follow this line of mathematical thinking, we would eventually understand why there are five cubes rather than, say, six or four. Indeed, we might then be in a position to make other, kindred objects by using the same underlying, structural principles.

This example illustrates an important point in science. By perceiving the internal structure of an object, you learn much about Nature; but by studying only its external form, especially in great detail and with high concern for nomenclature, you learn relatively little that is worthwhile.

Here we are going to describe the internal structure of DNA. The internal structure is the key to understanding how DNA works. We have tried to make everything as easy as spotting the structure of the five-fold cube. We may have to do a little mathematics, because mathematics tightens up the various relationships that we shall discover. There are a lot of things to learn about DNA, but we have tried to select the most important and general points. First we shall learn why DNA makes a *double helix*, as we saw in Chapter 1, instead of some other shape; and then we learn how the double helix is held together at its core.

You should recall from Chapter 1 that DNA is made of three things: phosphates, sugars, and bases, and that these components are linked together chemically in a particular way. Now the phosphates are very soluble in water. If farmers use too much phosphate on their land as fertilizer, some of the phosphate runs off into the nearby ditches and rivers, causing algae to grow wildly and so killing the fish. Phosphates are very soluble in water.

Sugars are also soluble in water. There would be no point in putting sugar into coffee or tea if it would not dissolve: if it just sat at the bottom of the cup it would not flavor the liquid. So sugars, as well as phosphates, dissolve in water.

But what about the bases? People have no intuitive feel for the four DNA (or RNA) bases: guanine, adenine, cytosine, and thymine (or uracil), because they don't recognize them in everyday life. It is easy to find out whether bases dissolve in water; just put some into a test-tube, add water and watch. You can buy adenine and uracil from any chemical company (they cost little) and use about 50 mg of each. When you do this experiment you find that neither adenine nor uracil dissolves in ordinary water. Yet further simple experiments show that adenine dissolves in weak acid and that uracil dissolves in ammonia, an alkali—but not vice versa.

Now we mentioned in Chapter 1 that the space in our cells which is not occupied by important components such as DNA, RNA, and enzymes is filled with water. This water is not at all acidic or alkaline. To use a technical term, it is at "neutral pH." It follows, therefore, that adenine and uracil (and indeed all of the bases listed above) will be practically insoluble in the aqueous environment of our bodies. Although we are not familiar with bases A, G, C, T, and U in everyday life, we are very familiar with other substances which will not dissolve in water, such as grease and oil. These are all "water-hating" or "hydrophobic" substances.

The insolubility of bases in water does not really pose a

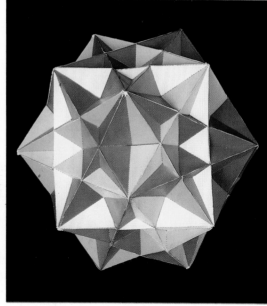

Figure 2.1 A puzzling object, in two views. With thanks to Ben Luisi. (From cut-out book by E. Jenkins, and M. Bear, (1985). Tarquin Publications, Diss, UK.)

problem for the cell, because these bases do become soluble in water once they are attached to a sugar and a phosphate to form a "nucleotide," which is the building-block of DNA or RNA (see Fig. 1.8). But this insolubility does place strong constraints on the overall conformation of any large DNA or RNA molecule in solution. For such a molecule to be stable in water at neutral pH, the bases will have to tuck themselves into the very center of some folded structure, so as to avoid the water; while the sugars and phosphates, both of which are soluble in water, will have to be on the outside. In fact, this is just what happens. If we take some measurements of the known structure of a DNA sugar–phosphate chain (determined by X-ray analysis, as described in Chapter 8), we see right away how DNA forms a spiral or helix on account of the low solubility in water of the bases. We can even do a first-order calculation to determine what kind of helix it makes.

We know from elementary chemistry that the distance between adjacent sugars or phosphates in the DNA chain is 6 Å (or 0.6 nm) in the usual case (Fig. 2.2). It cannot become much longer than 6.5 Å or shorter than 5.5 Å, or else the strong bonds

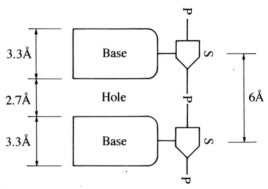

Figure 2.2 Two nucleotides in schematic form, showing key dimensions.

between the atoms will strain too much. The thickness of the flat part of a DNA base is 3.3 Å, and this distance cannot change much either, because the bases are chemically rigid with strong, inflexible bonds between the atoms. This leaves us with a "hole" of 2.7 Å between the bases, which some greasy object (and definitely not water) would have to fill, otherwise we will leave a vacuum. In short, the bases are attached to a sugar–phosphate chain that is twice as long as the thickness of the bases themselves.

How can we tuck these insoluble bases into the center of a DNA molecule where they can avoid water, and at the same time get rid of the "holes"? The most obvious form of DNA, as an assembly of two chains with the bases on the inside, would be a *ladder*. A segment of such a hypothetical ladder is drawn in Fig. 2.3. (A single chain of DNA could also fold back on itself to make a ladder, but this structure would be so similar to the one shown that it need not be considered separately.) In our ladder model, the two bases from opposite strands, joined in ways which we describe below, hold the phosphates 18 Å apart. Also, the two chains run in opposite directions, for reasons that we discuss below. All seems satisfactory in that respect, but we are still left with many "holes" between the bases within each strand. What can we do to remove them?

As shown in Fig. 2.4(a), one solution might be to skew the ladder strongly to one side. Once the sugar–phosphate chains tilt to an angle of about 30° from the horizontal, then the holes disappear, as shown. The key to the geometry is shown in Fig. 2.4(b), where we see a right-angled triangle that accurately describes the structure. The phosphates are connected along the hypotenuse of this triangle at a distance of 6 Å, while the bases

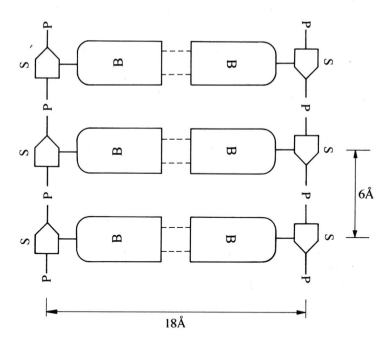

Figure 2.3 Part of a hypothetical DNA ladder, made by the cross-chain pairing of bases.

proceed upward by 3.3 Å along the right-hand side. The sideways-shift of the ladder per base-step, left to right across the paper, is therefore given by $x = \sqrt{(6^2-3.3^2)} = 5.0$ Å.

Our skew-ladder seems perfectly satisfactory as a way of

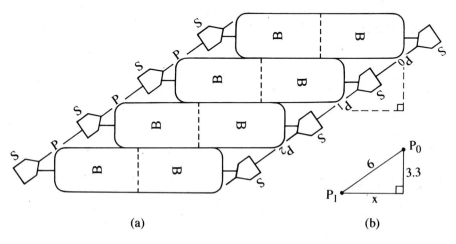

Figure 2.4 A skewed ladder, with no gaps between the paired bases. The plane geometry of this ladder is shown in part (b).

closing up the bases so as to exclude water. But it is not quite the same as the conformation which DNA adopts in Nature. As we have said in Chapter 1, DNA takes the form of a spiral or helix. In fact, the DNA double helix is nothing more than a highly twisted ladder. It provides another, slightly different way of solving the same problem: how to separate the bases by 3.3 Å while leaving the phosphates 6 Å apart. Figure 2.5(a) shows a simplified view of two base-pairs from our skew-ladder, while Fig. 2.5(b) shows that the bases can stack onto each other just as well, without gaps, if they twist about an imaginary vertical axis into the shape of a helix. The two chains climb from the horizontal at exactly the same angle as before—about 30°—but now they lie on the surface of a cylinder of diameter 18 Å; and the base-pairs within the helix are now arranged as in the treads of a spiral staircase.

Why does DNA prefer to form a helix rather than a skew-ladder? Our present model is too crude to answer this question convincingly. When we build a more accurate model which shows individual atoms, we find that the skew-ladder leads to many unacceptably close contacts between neighboring atoms, and so this model has to be abandoned. Nevertheless, the skew-ladder is a useful tool in thinking about the internal structure of DNA, because it has several important geometrical features of the real DNA helix, and yet lies in a plane and is easy to visualize.

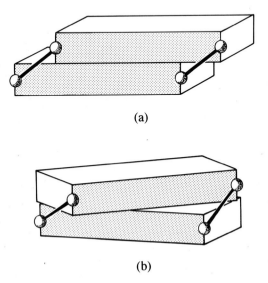

(a)

(b)

Figure 2.5 (a) Stacking of base-pairs as in the skewed ladder of Fig. 2.4. (b) Stacking of base-pairs by means of helical twist.

Now, when we go into three dimensions, and consider the shape of a DNA helix, the geometry is almost the same as in our skew-ladder above. We can take a series of the twisted, two-base-pair units shown in Fig. 2.5(b), and stack them on top of each other to get a proper, double-helical model of DNA. Figure 2.6(a) shows such a model schematically. Only the first two base-pairs are shown, but then we show all parts of the sugar–phosphate chains. These chains wrap as spirals around an imaginary cylindrical surface of radius 9 Å, and each sugar ring is represented by a dot. Figure 2.6(b) shows a side view of the cylinder for just one of the two sugar–phosphate chains. Here the phosphates, P_0, P_1, P_2, etc., are drawn as open circles, and the same lengths of 6.0 Å, 3.3 Å, and 5 Å that were found for our skew-ladder characterize the path of these phosphates through space. Finally, a top view along the vertical axis of the DNA cylinder is shown in Fig. 2.6(c). Again, for the sake of simplicity, only one chain is shown, and the phosphates along it are labeled P_0, P_1, ..., P_{10}. Each successive phosphate in this view lies 3.3 Å further away from us than the one before. The chain is shown with a break between P_{10} and P_0, because P_{11} lies directly behind P_0 in this view: it is 11×3.3 Å $= 36$ Å further away from us, when we look down into the plane of the paper.

Simple geometry enables us to calculate the angle θ (*theta*) by which each phosphate turns relative to its neighboring phosphate along the helix. As shown in Fig. 2.6(c), the distance $x = 5.0$ Å is the base of an isosceles triangle, whose vertex lies at the center of the cylinder at a distance of 9 Å from any phosphate. The value of θ can therefore be found by making a scale drawing, or else calculated as $2 \times \arcsin (2.5/9.0) = 32.3°$. Thus, each phosphate-to-phosphate rotation makes an angle of $32.3°/360° = 1/11$ part of a circle, and that is why we have put 11 phosphates in Fig. 2.6(c), to represent a complete turn of DNA.

This calculation, although relatively simple, tells us something which agrees closely with experiment: almost all DNA double helices have between 10 and 12 phosphates per turn of helix, within each strand. For example, the well-known "A" form of DNA (see Fig. 2.7) has 11 phosphates per turn, while the "B" form has 10, and the "Z" form has 12. These slight differences are significant in biology, and we shall discuss them later. But the crucial point here is that we have learned something important about the internal structure of the helix. Simply by studying the dimensions of the bases and of the sugar–phosphate chains, and knowing that the bases are insoluble in water (and

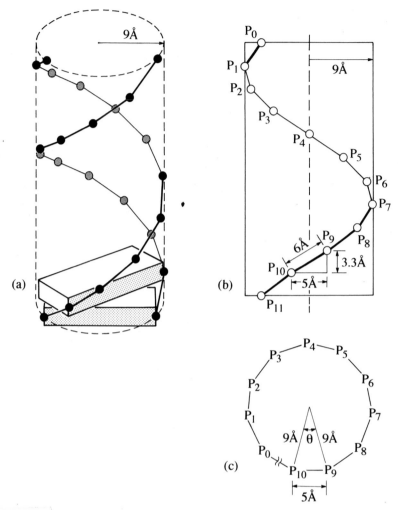

Sugar–phosphate chains wrapped helically around a cylinder: three views. In (a), sugar rings are drawn as shaded or filled circles, while phosphates are thin lines. In (b), phosphates are drawn as open circles, while sugars are thin lines. In (c), the view is down the long axis of the cylinder, looking along the dashed line in part (b).

so must stack directly onto each other), we have been able to determine that DNA will form a helix with about 11 phosphates per turn.

One point has been overlooked so far: how can we decide if our helix should be right-handed or left-handed? That is, as the phosphates spiral forward, should they go clockwise or counter-clockwise? There are, in fact, known examples of both right-handed and left-handed double helices: the "Z" form

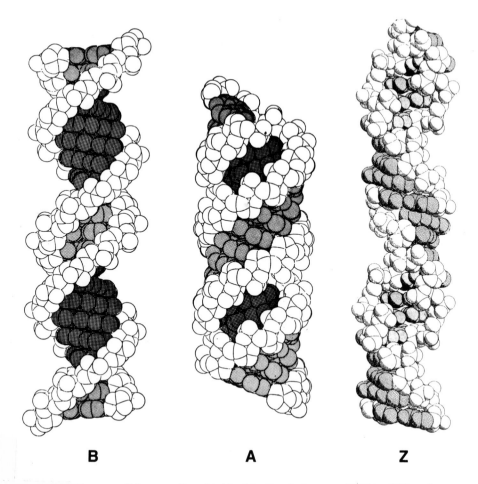

B   **A**   **Z**

Three well-known (but highly idealized) forms of DNA: "B" and "A" are right-handed with 10 and 11 phosphates per helical turn, respectively, while "Z" is left-handed with 12 phosphates per turn. Real right-handed DNA in solution averages about 10.5 phosphates per turn, or halfway between "B" and "A". Pictures of "A," "B" from C.J. Alden and S-H. Kim (1979) *Journal of Molecular Biology* **132**, 411–34. Picture of "Z" from H.R. Drew and R.E. Dickerson (1981) *Journal of Molecular Biology* **152**, 723–36 (with atoms shown somewhat smaller).

(Fig. 2.7) should perhaps be assigned −12 phosphates per turn, rather than +12, because it is left-handed rather than right-handed. It turns out that most DNA double helices are right-handed because of certain details of the chemical structure.

These details can be seen clearly when we build accurate, space-filling models of DNA: the atoms do not easily fit together if we try to build left-handed versions. Only very special combinations of DNA bases can become left-handed. Even then, the structures formed are so complicated and difficult to understand, that they will not be considered much in this book.

Before proceeding further, let us look at some realistic pictures of DNA that include all of the atoms. Figure 2.8(a) shows a space-filling model of a very small part of right-handed DNA, corresponding roughly to the fragment shown more crudely in Fig. 2.5(b), or to one small part of the "B" helix in Fig. 2.7. The atoms in this model are color-coded as hydrogen (white), carbon (black), nitrogen (blue), oxygen (red) and phosphorus (yellow). The bases are joined at the center in "pairs" (discussed below), while sugars and phosphates lie along the outside. This kind of model is widely used by scientists because it is so accurate. In fact, one could measure from the model a 3.3 Å separation of bases and a 6 Å separation of phosphates, and so check our previous calculation. Also, if you try to twist the base-pairs backward in this model, from right-handed to left-handed, the bonds between atoms fall apart.

Other more detailed chemical representations of just one DNA chain are shown in Fig. 2.8(b) and (c). There the atoms (or groups of atoms) are drawn as letters or balls: for example, P = phosphorus, O = oxygen, C = cytosine, A = adenine; while the bonds between the atoms are drawn as lines or sticks. The 5' to 3' direction of chains is also shown.

Let us now return to Fig. 2.7 to look more closely at the three forms of DNA shown there. Because we now know something about the internal structure of DNA, these models make sense to us. In all cases we can see that the bases fill compactly the centers of the double helices, where they successfully escape from contact with the surrounding water, and that the sugars and phosphates spiral around the outside of each helix at a rate of 11 plus or minus 1 phosphates per turn. The "Z" form also shows some jagged features in its sugar–phosphate chain that we shall not analyze here; they come from irregularities in the packing of bases on the inside of the helix. Using Fig. 2.8 as a guide, you can attempt to identify single bases, sugars, and phosphates in the pictures of Fig. 2.7, or even single atoms, if you have enough patience. It should be stressed that all of the pictures shown in Figs 2.7 and 2.8 are equally valid ways of representing the same thing; one drawing may be preferred over another, depending upon the level of detail at which you wish to visualize the DNA.

(a)

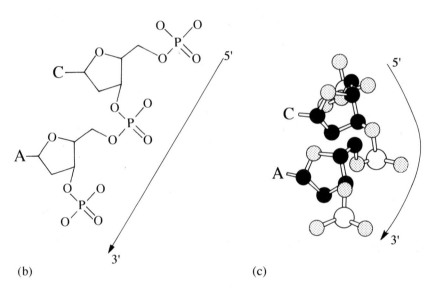

(b)   (c)

Figure 2.8 (a) Space-filling model of two stacked base-pairs, and their associated sugar–phosphate chains; (b) schematic layout of the right-hand chain, with bases cytosine (C) and adenine (A) shown only as letters; (c) three-dimensional version of (b), in same configuration as (a). Part (a) by courtesy of Louise Lockley.

We have assumed throughout our exposition that the two chains of a DNA double helix run in opposite or "antiparallel" directions (Figs 1.7, 2.3 and 2.4), because "parallel" double helices are not found in Nature. To a large extent, the interactions or pairings between bases at the core of the helix require that naturally occurring DNA be made from antiparallel chains. Some scientists have made short DNA molecules through chemical synthesis, in which the bases can only pair with one another when the chains run in a parallel sense: but these pairings are far less stable than when the chains run in the usual, antiparallel way, as found in Nature.

Another point before we go on. Scientists sometimes add to DNA a substance called ethidium bromide, which fluoresces a bright red-pink when ultraviolet light shines on it especially if it is bound to DNA. This provides a way of seeing exactly where the DNA is located in certain preparations, and so it constitutes an important tool for investigating DNA. How exactly does ethidium bromide stick to DNA? It is a greasy, mostly hydrophobic molecule, about the size of a base-pair. It can escape from contact with water by slipping between neighboring base-pairs along the chain, as shown in Fig. 2.9. Going back to Fig. 2.3, we must expect that the DNA will untwist locally to form an ordinary ladder before the ethidium bromide can fit; and indeed it turns out that DNA plus ethidium bromide make a largely untwisted ladder. This feature, combined with the fluorescence mentioned above, makes ethidium bromide an important tool for scientists who investigate DNA.

It is the *flexibility* of sugar–phosphate chains which allows them to change from a spiral to a straight ladder when ethidium bromide is added. This flexibility is of an unusual and indirect sort which arises from features shown in Fig. 2.10. There we see in (a) that the phosphate group is essentially a rigid tetrahedron, having a phosphorus atom at its center and one

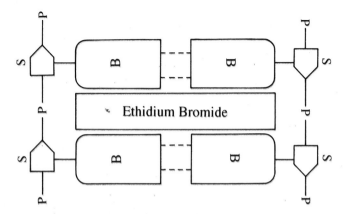

Figure 2.9 Ethidium bromide fills a gap between base-pairs (cf. Fig. 2.3).

oxygen at each vertex. Only when we go further along the chain from the phosphorus, as in (b), and attach two carbon atoms to two of the oxygens, can these carbon atoms swivel about the line of the phosphorus–oxygen bond. Imagine that your shoulder is the phosphorus, your elbow is the oxygen, and your hand is the carbon; obviously your hand and lower arm can swivel relative to your shoulder, but in DNA the angle at your elbow would be fixed. All parts of the DNA sugar–phosphate chain are rigid locally, but they have this kind of indirect flexibility over several bonds.

We have almost finished our survey of the basic principles that determine the structure of DNA. All we have to do now is to learn how the bases adhere to one another in the central core of the double helix. James Watson and Francis Crick solved this problem in 1953, by putting forward a set of rules for base-pairing. They said that the most stable base-pairs would be of the kind A–T or G–C, as shown in Fig. 2.11(a) and (b). One advantage of their scheme was that all four possible Watson–Crick base-pairs, A–T, T–A, G–C, and C–G, were of the same size, and hence could fit easily into the framework of a regular double helix. Another advantage was that it explained how the genes in DNA could be duplicated (or stably inherited) on cell division. Whenever a cell divides, and needs to duplicate its DNA, it can do so simply by splitting the DNA into two separate strands; then certain enzymes will come along and use each of these old strands as a "template" for the precise synthesis

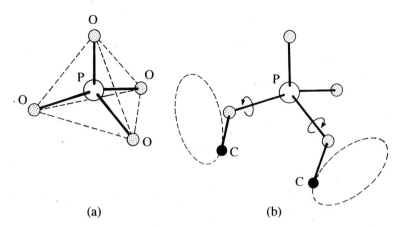

(a) (b)

Figure 2.10 (a) Tetrahedral arrangement of phosphate group: atom types are phosphorus (P) and oxygen (O); (b) freedom of rotation for adjacent links of the chain, with atom type carbon (C).

of a new strand, according to the Watson–Crick rules of base-pairing: A with T and G with C. (More will be said about this in Chapter 4.)

But what might be the physical basis of these Watson–Crick rules of pairing? They are based on the simple fact that, within any DNA base, there is a small surplus of negative electric charge on nitrogen and oxygen atoms where they are *not* attached to hydrogen, while there is a small surplus of positive charge on these same atoms where they *are* attached to hydrogen. Thus, consider the base-pair of adenine (A) and thymine (T), as shown in Fig. 2.11(a). Not counting the two nitrogens that are attached to sugars, there are three nitrogens on adenine and two oxygens on thymine that have a surplus of negative charge. On the other hand, one nitrogen on adenine and one nitrogen on thymine have a surplus positive charge. So all we have to do is to put the pluses and minuses together, thereby making the "hydrogen bonds" which are shown here as dotted lines.

There are, in principle, two ways of doing this, consistent with making a double helix. These are called the Watson–Crick and Hoogsteen base-pairs. The Watson–Crick A–T pair is shown in Fig. 2.11(a) while the Hoogsteen A–T pair is shown in Fig. 2.12(a). Both are roughly of equal stability: there are two hydrogen bonds which can be drawn between the bases in both cases. Note that the adenine base has to be rotated through 180° about the bond to the sugar in order to change between the two kinds of pairing—like rotating a tennis racquet from "rough" to "smooth." Note also that some of the interatomic connections within adenine and thymine rings are drawn as two lines, rather than as one: these are the "double bonds" which give the base rings both their flatness and their rigidity. Note also that the CH_3 or methyl group on the thymine ring would be absent in RNA, where the methyl-less base is called "uracil."

The history of these base-pairs is rather interesting. Watson and Crick found their A–T base-pair as part of a search for the double-helical structure of DNA, by playing with paper cut-outs of bases. (Some details of this are given later, in Chapter 8.) Karst Hoogsteen tried 10 years later to confirm the Watson–Crick pair for adenine and thymine by heating up a solution of these two bases and letting it cool slowly in order to make a crystal; but he found instead a different kind of base-pair in his crystal.

Watson–Crick and Hoogsteen pairs for guanine (G) and

Figure 2.11 Watson–Crick base-pairs, showing hydrogen bonding: (a) A–T; (b) G–C. Atoms which are not labeled are carbon, while other atom types are hydrogen (H), nitrogen (N), and oxygen (O). Not all hydrogens are shown. Hydrogen bonds are represented by dotted lines. Symbols +, − here represent partial electric charges of about $\frac{1}{3}$ electron or proton, which are typical for hydrogen bonds.

cytosine (C) can also be drawn as shown in Fig. 2.11(b) and 2.12(b). There are two things to note here. First, the Watson–Crick guanine–cytosine pair has three hydrogen bonds, rather than two as for adenine–thymine; so DNA double helices with guanine and cytosine should be more stable than those with adenine and thymine. Secondly, the Hoogsteen guanine–cytosine pair is only stable at low pH, since one of the nitrogens on cytosine must be protonated (i.e. be attached to a hydrogen) for this structure to form. The midpoint for protonation is pH 5, or slightly more acidic than the normal pH 7–8 found in cells. That is the main reason why practically all DNA double helices contain Watson–Crick rather than Hoogsteen pairs; the Hoogsteen G–C pair is *not stable at neutral pH*. An additional reason is that a

Figure 2.12 Hoogsteen base-pairs, showing hydrogen bonding: (a) A–T; (b) G–C. Symbols +, − are partial charges as in Fig. 2.11. The Hoogsteen G–C pair is stable only at mildly acidic pH (about 4–5), because it requires protonation of a cytosine nitrogen, i.e. the adding of a hydrogen to it.

Watson–Crick G–C pair has more hydrogen bonds than a Hoogsteen G–C pair—three versus two. In principle, DNA containing only A and T bases should be relatively stable in either Watson–Crick or Hoogsteen forms, but few purely Hoogsteen-paired double helices have yet been detected with certainty. Also, it seems unlikely that both Watson–Crick and Hoogsteen pairs could be easily accommodated in close proximity within the same double helix, because the two kinds of base-pair are shaped differently.

Until recently it would have been enough to learn just about Watson–Crick base-pairs. But now it is important to learn about Hoogsteen pairs as well, because such pairs show up occasionally in complexes of DNA with anticancer drugs, and also in triple helices where a third strand of DNA joins the first two. The base "triplet" then contains both Watson–Crick and Hoogsteen pairs: try to draw a triplet with one adenine and two thymines. Two scientists in California, Scott Strobel and Peter Dervan, have actually been able to cut a yeast chromosome cleanly in half, by designing a third strand of DNA that binds only to one specific double-helical sequence within the whole chromosome, and then by attaching an iron atom to this third strand so as to "rust" (or oxidize) the DNA into pieces.

Why can't there be other stable base-pairs, such as G with A, or C with T? Some of these are ruled out by the difficulty of making two or more hydrogen bonds. But others, such as guanine (G) with thymine (T) or uracil (U), as shown in Fig. 2.13, are not excluded for that reason. The hydrogen bonding produces a pair with similar overall shape to those in Fig. 2.11. In fact, a guanine–uracil pair is perfectly stable, and is used when transfer RNA—the molecule that carries amino acids for protein synthesis—binds to messenger RNA on the ribosome. You may recall from Chapter 1 that DNA makes a copy of itself in the form of RNA, and that this copy, called "messenger RNA," travels outside the nucleus to the protein-making machinery or "ribosome." There it becomes attached to a series of different transfer RNA molecules, one of which is shown in Fig. 2.14. Each kind of transfer RNA contains three specific bases at one end, and one specific amino acid at the other. If the three specific bases on transfer RNA can form base-pairs with three neighboring bases in the messenger RNA, then the amino acid

Figure 2.13 Pairing of G with T allows two good hydrogen bonds, plus an overall shape similar to that of the Watson–Crick pairs shown in Fig. 2.11(a) and (b). A closely related base-pair, that of G with U (uracil), is used routinely to specify amino acids for the synthesis of proteins, as shown in Fig. 2.14.

being carried by the transfer RNA is added to a growing protein chain.

Now there are 4 × 4 × 4 = 64 possible combinations of three bases in the messenger RNA, and yet sometimes there are only 25–30 different kinds of transfer RNA molecule in a cell. So several different sets of three bases in the messenger RNA must have to "share" a transfer RNA, in much the same way that two birds will share the same nest when there are not enough trees. In fact, the transfer RNA is flexible, and will accept either of two possible base-pairs in the third position. An example is shown in Fig. 2.14: here either an RNA sequence GCU with an "unusual" G–U pair, or else a sequence GCC, with the "normal" G–C pair, serves to determine one particular amino acid. A guanine–uracil pair is not the only unusual one to be used in this way; there are several others.

That brings us to a final question: if all of these different base-pairings are stable enough to be used by transfer RNA when it links with messenger RNA, as in Fig. 2.14, why aren't they found in DNA? If they were used every time a cell divides, the sequence of bases in the DNA could change drastically. Yet this evidently does not happen, since it is now known that even a single G–C to G–T change at a critical place in the DNA program could cause cancer. In fact, Nature uses proteins called "proofreading enzymes" to prevent the occurrence of slight

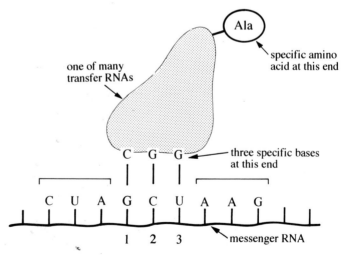

Figure 2.14 The transfer RNA molecule that carries alanine can recognize its preferred triplet of bases, on a messenger RNA chain, by using either a G–U or a G–C base-pair in position 3.

changes in sequence when DNA replicates. The enzymes that copy DNA to DNA, or DNA to RNA, are indeed very clever. They can sense at several stages during synthesis whether anything is going wrong; for example, if they have added or are about to add the wrong base, according to the Watson–Crick rules of pairing. Also, there are "repair" enzymes that go around correcting occasional mistakes of copying, or "mismatches". Thus, Nature goes to great lengths to avoid errors in the copying of DNA, even though the atoms in the DNA structure are actually quite tolerant of mismatch pairings. These enzymes are extremely efficient in doing their job, yet no one knows exactly how they work.

In summary, we have learned in this chapter how the insolubility of bases in water provides the driving force for DNA to form a double helix; and how the geometry at the core of the helix depends on subtle interactions between partial electrical charges on the bases. These subtle interactions alone are not sufficient for accurate copying of DNA from generation to generation, and so cells contain many enzymes that improve upon the efficiency of copying. Yet unusual base-pairs are a fact of life at other places in the cell, such as when proteins are made via the binding of transfer RNA to messenger RNA; so one has to be aware of all these subtle possibilities, in order to appreciate the internal structure of nucleic acids as they act in biology.

Further Reading

Arnott, S. (1970) The geometry of nucleic acids. *Progress in Biophysics and Molecular Biology* **21**, 265–319. A summary of early X-ray studies on fibrous samples of DNA.

Dickerson, R.E., Drew, H.R., Conner, B.N., Wing, R.M., Fratini, A.V., and Kopka, M.L. (1982) The anatomy of A-, B- and Z-DNA. *Science* **216**, 475–85. A summary of the first work from X-ray studies of DNA in single crystals.

Drew, H.R., McCall, M.J., and Calladine, C.R. (1988) Recent studies of DNA in the crystal. *Annual Reviews of Cell Biology* **4**, 1–20; or (1990) New approaches to DNA in the crystal and in solution. In *DNA Topology and Its Biological Effects* (eds N.R. Cozzarelli and J.C. Wang), ch. 1, pp. 1–56, Cold Spring Harbor Laboratory Press. Up-to-date reviews of recent X-ray and solution work on DNA structure, and on the role of DNA structure in biology.

Saenger, W. (1984) *Principles of Nucleic Acid Structure*. Springer-Verlag,

New York. A comprehensive review of the published literature on DNA structure up to 1982.

Bibliography

Hoogsteen, K. (1959) Structure of a crystal containing a hydrogen-bonded complex of 1-methylthymine and 9-methyladenine. *Acta Crystallographica* **12**, 822–3. Discovery of an alternative pairing-scheme for A with T.

Strobel, S.A. and Dervan, P.B. (1990) Site-specific cleavage of a yeast chromosome by oligonucleotide-directed triple-helix formation. *Science* **249**, 73–5. Use of a triple helix with iron (Fe) attached, to "rust" a yeast chromosomal DNA molecule into two pieces.

Watson, J.D. and Crick, F.H.C. (1953) A structure for deoxyribose nucleic acid. *Nature* **171**, 737–8. First proposal of the rules for base pairing of A with T and G with C, and of a two-stranded, double-helical model for DNA.

Exercises

2.1a On planet P all living things are found to contain a DNA-like double-helical molecule just like that found on Earth, except that the molecule on planet P consists only of the two nucleotides A and G. Thus, A–G is the only scheme for base-pairing, and the large A–G base-pairs impart a separation of 20 Å to the two sugar–phosphate chains. By adapting the calculations shown in Figs 2.4 and 2.6, estimate the angle of helical twist and the number of base-pairs per helical turn for the special DNA on planet P.

b On planet Q the DNA molecule contains four nucleotides A, G, C, and T, just as on Earth; but the sugar–phosphate chain is found to be 7 Å long between phosphates (compared with 6 Å on Earth) on account of an extra carbon atom in each nucleotide unit. Given that the bases pair in Watson–Crick style, how many base pairs do you expect per double-helical turn in the DNA on planet Q?

c On planet R the genetic molecule is exactly like terrestrial DNA, except that the oceans on planet R are slightly acidic (pH 4), such that the base-pairings A–T and G–C are mostly in accordance with the Hoogsteen scheme (Fig. 2.12). In consequence, the sugar–phosphate chains spiral about an imaginary cylinder of diameter 16 Å. Estimate the number of base pairs per helical turn of DNA on planet R.

2.2 Idealized "B" form DNA has a helical twist of 36° and "rise" of 3.3 Å, in the axial direction, per base-pair step (see Fig. 2.7). When a molecule of the intercalating drug ethidium bromide inserts itself into a step of DNA in the manner of Fig. 2.9, it increases the length of the DNA by 3.3 Å, and at the same time reduces the helical twist at the step by 26°, i.e. from 36° to 10°.

Find the overall length of a 100-base-pair segment of "B" form DNA, and the total number of helical turns:

  a with no ethidium bromide;
  b with one ethidium bromide molecule for every 10 base-pairs;
  c with one ethidium bromide molecule for every 2 base-pairs.

(Case **c** corresponds to the largest possible uptake of ethidium bromide by DNA.)

2.3 In some circumstances, DNA can make a triple helix by forming planar hydrogen-bonded base triplets in place of the usual base-pairs. Construct such a triplet from one A and two Ts in two different ways, with each thymine T connected by two hydrogen bonds to the adenine A.

  a Begin with the A–T pair of Fig. 2.11(a), and add a second T to the "unoccupied" upper edge of the A, by moving the first T base around in the plane of the paper.
  b Begin with the usual A–T pair, as before; but now obtain the second T by flipping the first T over onto its other face before moving it around.

Which of the new pairings is equivalent to the Hoogsteen arrangement of Figure 2.12(a)?

(Hint: Work with a copy of Fig. 2.11(a), and use tracing paper—either way up—for the second T. Note that either oxygen of the thymine can act as a hydrogen-bond acceptor.)

2.4 Investigate possible G–A base-pairings as follows: begin with the G base from Fig. 2.13, along with its hydrogen-bonding scheme; and use A bases with the hydrogen-bonding schemes of Figs 2.11(a) and 2.12(a), but "flipped over" in each case. Work on tracing paper. Do the two different G–A pairings differ much in their sugar–sugar distances?

2.5 The three-base code in a DNA or messenger RNA chain (see Table 1.1) is read with the assistance of various transfer RNA molecules, as shown schematically in Fig. 2.14. There, one particular transfer RNA molecule is shown recognizing the messenger RNA sequence GCU (corresponding to GCT in DNA); but the same transfer RNA can also recognize the sequence GCC, by using a standard G–C pair rather than an unusual G–U in position 3 of the message.

a   Suppose that a particular cell has 32 different kinds of transfer RNA molecule, each with one of 20 possible amino acids attached at its distal end. On average, how many different sets of base-triplets in the messenger RNA chain would each transfer RNA have to recognize, in order to make all the proteins necessary for cell growth?

b   What is the smallest number of different transfer RNA molecules that the cell could use and still be viable? How many triplets would any particular transfer RNA molecule then have to recognize, on average?

CHAPTER 3

Different Kinds of Double Helix

In the previous chapter we learned two things: why DNA forms a double helix, and how the bases interact with one another at the oily, water-insoluble core of the helix to form "base-pairs." The driving force for helix formation was shown to be the need for the bases to escape from water, by joining with other bases at the core of the helix. Yet they cannot stack directly on top of one another while doing so: rather, they must twist around slightly, because they are attached to sugar–phosphate chains that are twice as long for each base as the thickness of the base itself.

All of these points are what might be called "first-order" influences on the structure of DNA. Now you might be hoping that you will not need to learn about "second-order" effects: you can perhaps manage by just knowing the first-order effects. That is a forlorn hope, however, because you will have to learn about the second-order effects before you can understand many of the roles of DNA in biology: for example, how promoters work, how DNA wraps into chromosomes, and even how DNA binds to the "repressor" proteins which influence how well promoters work. As you may recall from Chapters 1 and 2, the DNA within any gene makes an RNA copy of itself which then goes on to make protein; a "promoter" is a short region of DNA near the gene that tells the cell how many RNA copies to make, and hence how much protein.

We shall limit ourselves to three themes in our study of second-order effects on the structure of DNA: first, how the bases undergo "propeller twist" to make sure that as much as possible of their oily, flat surfaces escape from contact with water; second, how the base-pairs stack on one another in particular ways that depend upon the ordering or sequence of bases; and third, how certain small, subtle motions of the base-pairs can accumulate over a series of such base-pairs, to make different kinds of double helix.

We said in the last chapter that the bases form ordered pairs at the core of the helix, leaving sugars and phosphates on the outside; and that each base-pair twists with respect to its neighbor by about 32°, as in the treads of a spiral staircase. The sense of this rotation is right-handed, or clockwise going forward as shown in Fig. 3.1: this is the same helical sense as in an ordinary corkscrew.

Almost always, the base-pairs are of a Watson–Crick kind, joining guanine (G) with cytosine (C), and adenine (A) with thymine (T): see Fig. 3.2. Bases A and G are called "purines," and they are bigger than bases C and T, which are called "pyrimidines;" yet the overall size of the base-pair is roughly the same in all four possible arrangements. The apparent simplicity of these arrangements once led scientists to conclude that the base sequence of DNA could not influence its three-dimensional structure, because all four kinds of base-pair could slot into a perfectly uniform double-helical "staircase." This conclusion was not based on any firm evidence, however, and it has proved to be incorrect. For example, typical angles of base-pair twist in real DNA molecules, as determined from the many high-resolution maps of DNA structure collected since 1980, range from 28° to 40° about a mean of 34°. The mean value of 34° is close to our prediction of 32° from the simple theory presented in Chapter 2; yet our first-order theory did not predict anything about a variation in twist away from 32° to other values. It certainly did not predict a broad range of twist from 28° to 40°, as is typically observed. Evidently, we must develop a theory that includes these second-order effects before we can claim to have any real understanding of the structure of DNA.

Our starting point for the second-order theory is rather subtle: it seems that because of the substantial twist between adjacent

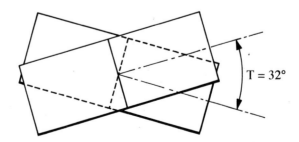

Figure 3.1 Two base-pairs with 32° of right-handed helical twist: the minor-groove edges are drawn with heavy shading, as in Fig. 3.5.

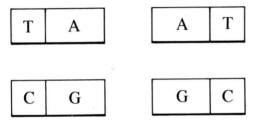

Figure 3.2 Four possible base-pairs of the Watson–Crick type, each of which joins a large purine (A, G) to a small pyrimidine (T, C), and in the same perspective as Fig. 3.1.

base-pairs, less than the entire surface of any base-pair can escape contact with water. Thus, only the central overlapping portions of the base surfaces in Fig. 3.1 are protected from water, while the four overhanging triangular portions are not protected. When we view the two right-hand bases in Fig. 3.1 from the perspective of the right-hand margin of the page, edge-on in the plane of the paper, we see the arrangement shown in Fig. 3.3(a). Taken together, Figs 3.1 and 3.3(a) show that the overlap of consecutive pairs is good in the interior of the stack, but only poor in the outer regions, which are exposed to the water.

What can we do to improve this situation? One solution would be to rotate each of the bases shown in Fig. 3.3(a) in a clockwise sense about its long axis, which points down into the plane of the paper. Such a motion is shown in Fig. 3.3(b), where each base rotates slightly about its end-centerpoint, which is a black dot in the diagram. Stacking is improved, since the water is now excluded from a larger fraction of the surfaces of the two bases. If this motion isn't clear to you, stretch out both of your arms in front of you, with your right hand above your left, both

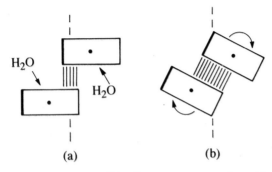

(a)   (b)

Figure 3.3 Propeller twist, as in (b), allows greater overlap of bases within the same strand and reduces the area of contact between the bases and water.

hands horizontal, and with the tips of the middle fingers vertically above each other and about 5 cm apart. Then move your right hand by about 5 cm in a left-to-right sense, so that it no longer lies directly over your left. Finally, rotate both hands clockwise about the wrists by around 20°; and you will find that your two hands "cover" each other much better than they did before.

Now the two bases on the left-hand side of Fig. 3.1 can likewise rotate about their long axes to exclude water, but their sense of rotation must be counterclockwise, when viewed from the perspective of Fig. 3.3, in order to achieve the same result. All four bases from Fig. 3.1 are shown together in Fig. 3.4. In this picture, each base-pair looks somewhat like an old-fashioned airplane propeller, since the left-hand and right-hand bases twist in opposite directions. Hence, the overall motion is called "propeller twist." Its sense is that of a left-handed screw as one goes forward along the pair from one base to its partner. Propeller twist obviously distorts the hydrogen bonds that hold the two bases together, which are shown schematically by two lines in Fig. 3.4; but these weak bonds can accept some distortion of that kind, provided the degree of propeller twist is not too great. Typically, one sees 10° to 20° of propeller twist in real DNA structures.

Note also that we have shaded one edge of each base in Fig. 3.4, and have labeled one of these with the term "minor." There is a convention to call one side of a base-pair the "minor-groove side," and the other the "major-groove side." Where do these names come from? As shown in Fig. 3.5(a), the two sugars to which a base-pair are attached lie closer to one side of the base-pair than the other. The edge which lies closer to an

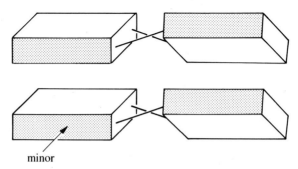

minor

Figure 3.4 Propeller-twisted base-pairs. Note how the hydrogen bonds between bases are distorted by this motion, yet remain intact. The minor-groove edges of the bases are shaded.

imaginary line drawn between the two sugars is called the "minor-groove side," while the other edge is called the "major-groove side." By convention, we shall always shade the minor-groove side of a base-pair in our drawings, as indicated in Fig. 3.5(b).

Why do we use the term "groove" in this labeling convention? Early structural models of DNA showed a cylinder with two hollow, spiral grooves lying between the two sugar–phosphate chains. One of the grooves—"minor"—was smaller than the other. You can see these two grooves in the "B" model of Fig. 2.7. But for some of the helices which we shall study in this chapter, it turns out that the so-called minor groove is actually as large as or larger than the so-called major groove: for example, see the "A" model of Fig. 2.7. We can get around this difficulty by talking instead about the minor-groove *side* of the base-pairs themselves.

Of course, we can also view the base-pairs from the major-groove side. An example of this is shown in Fig. 3.6, where we see that the near edge, labeled "major," remains unshaded. Furthermore, in this drawing we can see that the two bases A and T within either pair are of unequal size. In general, as we have said, purines A and G are larger than pyrimidines T and C. In several previous drawings, for example Figs 3.1 and 3.4, we have omitted this feature for the sake of clarity; but when we start to consider the interactions of real DNA bases such as those shown in Fig. 3.6, we have to make our drawings more

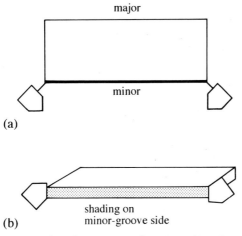

(a)

(b)   shading on
minor-groove side

Figure 3.5 Two views of a base-pair, showing directions of the sugar–phosphate chains, just as in Figs 2.3 and 2.4. By convention, the minor-groove edges are shaded.

accurate. Figure 3.6 also shows the detail of a hydrogen bond
(N–H . . . O) between adjacent base-pairs, from adenine in the
lower pair to thymine in the upper. Such a hydrogen bond
might be expected to increase the propeller twist, because the
distance is right for it to form only when the base-pairs are
highly twisted along their long axes. Indeed, experiments show
that regions of DNA with all adenine bases on one strand and
all thymine bases on the other do have an unusually high
propeller twist, of about 20–30° as against 10–20° for other
sequences.

You can imagine that if we were to study all kinds of two-
base-pair arrangements in DNA, or indeed all kinds of three-
base-pair arrangements, we should find a lot of unexpected but
important contacts between the bases. For that reason it is very
hard to understand, on an atomic scale, the behavior of any
long DNA molecule such as those found in biological systems.
We need some sort of simplified description of base-pair
arrangements in DNA, at less than an atomic level of detail, if
we wish to understand the many roles of DNA in biology.

The simplification which we make at this point is one which
we have already used in some drawings: we construct an
imaginary, flat plane that coincides as well as possible with the
twisted surface of any base-pair. In other words, we shall
pretend that the base-pairs shown in Figs 3.4 and 3.6 have no
propeller twist, and treat them as rigid, rectangular blocks such
as those shown in Figs 3.5 and 3.7. Previously, of course, we
have said how important it is to build propeller twist into a
base-pair; but now it seems that we are going to ignore this
twist entirely in a simplified model! You must simply remember
that the propeller twist is always *there*, but that it is not always

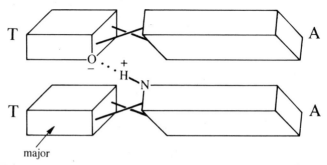

Propeller-twisted A–T pairs, showing an additional hydrogen bond
between the base-pairs in the major groove (as proposed originally by Hillary
Nelson).

shown in the diagrams. This is a bit like drawing a watch without showing the gears inside: they are there, but not visible. For some purposes we can understand things more clearly by thinking about a deliberately simple representation of a base-pair.

Our simplified model is based on the work of the famous Swiss mathematician, Leonard Euler (1707–83). He explained that if you have two rigid objects such as the rectangular blocks shown in Fig. 3.7—here representing two base-pairs—and you want to describe the position of one block relative to the other, you will need to use six variables or "degrees of freedom": three translations and three rotations. A translation[1] is a change of position without any rotation: imagine moving a cardboard box from one place in a room to another, so that every face of the box always moves parallel to itself. The position of the box in the room can be specified completely by the values of three "coordinates," say x,y,z, measured from a suitably chosen "origin." A rotation, on the other hand, involves a change of angle without a change in position: for example, you could pick up a cube from a table, turn it through 90°, and then put it back on the table in the same position as before.

It is easy to understand why you need six variables to describe completely the position of one solid block relative to another. Suppose that the upper base-pair or block shown in Fig. 3.7 were temporarily replaced by a *point*: then you would only need three translation coordinates, x,y,z, to say where this point might be located relative to the lower block. Next let us build an upper block around the point, and ask how many kinds of rotation we need to orient the upper block relative to the lower block. The answer is exactly three, as indicated by the three

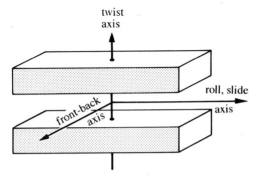

Figure 3.7 Local reference axes for an individual base-pair step, following the mathematics of Leonard Euler.

axes of rotation drawn in Fig. 3.7. We call these the "twist," "front–back," and "roll–slide" axes for reasons that will become clear soon.

So it seems that we only need six numbers to describe the local configuration of any two neighboring base-pairs, from a mathematical point of view. That is not too bad. But when we look at real DNA structures, the situation becomes even more favorable. To a first approximation, only three of Euler's six possible degrees of freedom are actually mobilized in real DNA double helices. This fortunate simplification comes about because the base-pairs are attached, as we recall from previous chapters, to sugars and phosphates which limit their range of maneuver in certain directions, notably along the front–back axis of Fig. 3.7. Also, the base-pairs cannot be separated along their vertical twist axis without introducing water or a vacuum between them.

Thus, in practice, we need only three variables to describe the motions of any base-pair relative to its neighbor. Two of these are rotations, and one is a translation. All three are shown separately in Fig. 3.8. The first rotation, or "twist," (Fig. 3.8(a)) is the same twist which we estimated as about 32° in Chapter 2, and which we showed in Fig. 3.1. We shall now give it a rigorous definition: it corresponds to a rotation about the local twist axis that runs vertically through, or near, the centers of any two neighboring base-pairs, as shown in Fig. 3.7. Note that if the DNA happens to be curved, or if the base-pairs stack on one another locally such that they do not advance directly along the overall helix axis, then the *local* twist axis shown in Fig. 3.7 may not coincide with a *global* twist axis for the whole molecule, averaged over many steps; and the twist angles measured about these two different axes may also be slightly different. So it is important to define what kind of twist we mean, when we talk about the "twist" of DNA.

The second kind of rotation, or "roll" (Fig. 3.8(b)) describes the rolling-open of base-pairs along their long axes. Angles of roll vary from +20° to −10° in the usual DNA structures. By convention, we say that the roll is positive if base-pairs open up towards the minor-groove side, as shown in the diagram. Actually, the surfaces of individual bases do not come apart from one another very much in the roll motion: they only appear to do so because we have drawn the two base-pairs as uniform, rigid blocks. This simple kind of drawing conceals details, in much the same way that the cover of a watch conceals the gears.

The last commonly observed kind of motion is a translation, or "slide" (Fig. 3.8(c)). It describes the relative sliding of

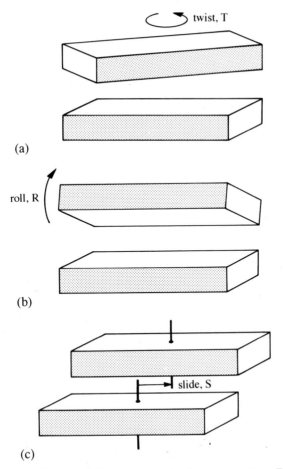

twist, T

(a)

roll, R

(b)

slide, S

(c)

Figure 3.8 Twist, roll, and slide motions at a base-pair step. Each drawing defines the positive sense of twist, roll, or slide, as used in this book.

neighboring base-pairs along their long axes. Slide is defined as positive if the upper pair goes further to the right than the lower pair, when we look at the minor-groove edges. Typically, values of slide range from +2Å to −1Å in real DNA; the sugar–phosphate chains do not easily allow any further motion. Still, these sugar–phosphate chains do allow a great deal more flexibility about the roll–slide axis, as epitomized by "roll" and "slide," than they do about the front–back axis shown in Fig. 3.7.

In summary, there are three significant relative motions of the base-pairs at any base-pair step. They are called roll, slide, and twist, and they may be abbreviated to R, S, and T, respectively. The positive sense of each is shown in Fig. 3.8(a), (b), and (c).[2] You can't forget the series R, S, and T, although you have to remember which letter stands for which degree of freedom.

The propeller-twist motion, which we were talking about earlier, is a property of a single base-pair, and not of two base-pairs that lie over one another. Thus, we can imagine two short pieces of DNA that have different values of propeller twist but the same values of roll, slide, and twist. As shown in Fig. 3.9, if we stack two base-pairs directly on top of one another, at zero slide and zero twist for the sake of simplicity, then the roll remains zero no matter how much propeller twist we add equally to both pairs. At the level of the "rigid block" drawings shown in Figs 3.7 and 3.8, both parts (a) and (b) of Fig. 3.9 would look identical, despite the change in propeller twist on going from (a) to (b). Bases on the right-hand strand tilt upward along their minor-groove edges by 10°, and those on the left-hand strand tilt downward by 10°; nevertheless, the two effects cancel when we calculate a mean plane for the entire base-pair.

Drawing base-pairs as rigid blocks, and then identifying roll, slide, and twist motions at the steps between base-pairs may seem to be a strange way of thinking about DNA; but it is the most successful way so far devised. Therefore let us continue, and see how the three parameters roll, slide, and twist might be useful. Even though we have pushed the idea of propeller

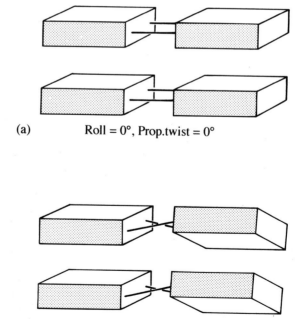

(a)   Roll = 0°, Prop.twist = 0°

(b)   Roll = 0°, Prop.twist = 20°

Figure 3.9 Propeller twist need not alter roll. The roll angle remains zero in part (b), because the mean planes of the base-pairs remain parallel.

twist out of the limelight for the moment, we must not forget about it, because propeller twist is partly responsible for certain relations between R, S, and T which are found in real DNA, as we shall see shortly.

Previously, in Fig. 3.6, we noted one important feature of DNA structure, which was a hydrogen bond between neighboring base-pairs in sequences of the kind AA or TT. Another important feature of DNA structure can be found at steps which we shall describe as "pyrimidine–purine." Figure 3.10 shows an example of such a step. You should have learned from the many pictures shown so far that the two sugar–phosphate chains of DNA run in opposite directions. In Fig. 3.10, these sugar–phosphate chains are not shown, but their directions are marked by arrows. Suppose in a particular step that the upper base-pair is C–G and the lower one is A–T, in going from right-to-left across the page. Then the entire step can be described as "CA/TG": coming down along the right-hand chain we have the sequence CA, and going back up along the left-hand chain we have TG. The "slash" between CA and TG indicates the "jump" from one chain to the other. If you were to turn the book through 180° and use the same procedure for naming this particular step, you would get "TG/CA" instead of "CA/TG." It doesn't matter which term you use: both describe the same collection of atoms in space.

The drawing in Fig. 3.10 is intended to represent all possible pyrimidine–purine steps, which we may now list as TG/CA (= CA/TG), CG/CG and TA/TA. These steps are seen commonly with two different kinds of base-to-base overlap, as shown schematically in Figs 3.10 and 3.11.

In Fig. 3.10, the large purine bases G or A slide away from each other by −1 Å, in order to avoid too-close contact at the point shown by the "star" in the picture. If the base-pairs were not propeller twisted, then contact of purines at the star would not be a serious problem, and the slide could be zero. Note that the roll angle is zero, since all bases remain parallel to their neighbors on the same strand.

In Fig. 3.11, the large purine bases slide on top of one another by +2 Å. Now the roll angle becomes large and positive, +20°, since the small pyrimidine bases must be inclined with respect to the large purine bases in either strand by +20°, in order to maintain the 20° of propeller twist. Apparently, the close stacking of pyrimidine on purine within either strand of real DNA is less significant than is the maintenance of high propeller twist.

There are two important points about Figs 3.10 and 3.11. First,

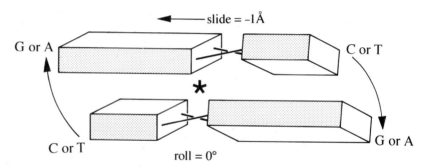

Figure 3.10 A pyrimidine–purine step with zero roll: negative slide is needed to avoid a steric clash at *, if the base-pairs have propeller twist.

values of slide lying between the extremes of -1 and $+2$ Å are expected to be less stable than the ones shown, because at intermediate values (say 0 to $+1$ Å) the two large purine bases will neither avoid one another fully, nor stack firmly on top of one another. One expects, therefore, that pyrimidine–purine steps in DNA should be weakly "bistable," that is, capable of adopting two discrete conformations but not a continuous range.[3] Second, by comparing the two structures shown in Figs 3.10 and 3.11, we can see that roll changes as we change the slide, on account of propeller twist. Thus, slide $= -1$ Å gives roll $= 0°$, while slide $= +2$ Å gives roll $= +20°$. This is very much like the motion of a bolt-action rifle: the bolt "rolls" as it "slides" forward to place a bullet in the chamber.

Would other sequences in DNA show these same characteristics, or would they behave differently? For example, would we expect the AA/TT step in Fig. 3.6 to be bistable, despite the stabilizing influence of an extra hydrogen bond in the major groove? And if the AA/TT step did slide forward to $+2$ Å, would we expect the roll angle to increase to $+20°$ when both purines lie on the same strand?

The answer, as you might expect, is that only the pyrimidine–purine steps show both bistability and roll–slide correlation to such a strong degree. The best way to study this problem is to plot roll and slide values for many of the base-pair steps that have been determined by X-ray diffraction analysis, and then look for significant features. Such a plot is shown in Fig. 3.12. Even at first glance, one can see that most of the data lie on a broad but sloping band from lower left to upper right, or in other words, from low roll–low slide to high roll–high slide. The coupling of roll with slide is a direct consequence of propeller twist within individual base-pairs.

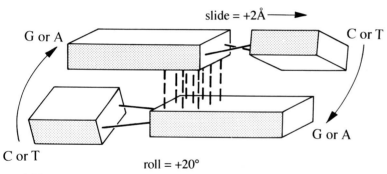

Figure 3.11 A pyrimidine–purine step in an alternative configuration, having positive roll and slide, due to the cross-chain stacking of purines.

Pyrimidine–purine steps, circled by dashed lines in the figure, lie solely in two zones that correspond to the two arrangements shown schematically in Figs 3.10 and 3.11. Other steps follow the broad, nearly continuous distribution of points from lower left to upper right, but they do not cluster at its ends to such a significant degree. Also drawn on the plot of Fig. 3.12 is an almost-vertical dashed line, marking an approximate boundary between the well-known "B" and "A" types of double helix (see Fig. 2.7). There is no sense in worrying too much about whether a given double helix belongs to the "B" or to the "A" type, because the classification into one of these two categories is somewhat arbitrary and often ambiguous.

Before going further, let us give a simple description of what we have been saying about base-step stacking arrangements in this chapter. Suppose the base-pairs in DNA were flat like pancakes, with no propeller twist; then roll and slide could vary independently of one another, subject to the constraints of attachment to the sugar–phosphate chains. But since the base-pairs are usually not flat, due to their substantial propeller twist, then they behave like step-twisted plates: when you slide one plate down the long axis of its neighbor, the two plates do not remain perfectly parallel but roll apart slightly.

Looking again at the pictures in Figs 3.10 and 3.11, which describe the favored conformations of a pyrimidine–purine step, one might imagine a similar kind of motion for a purine–pyrimidine step such as GC/GC, GT/AC (= AC/GT) or AT/AT. In this case, the large purines from opposite strands would either overlap at low slide to yield roll = −20°, or move apart at high slide, avoiding each other at a "star" on the major-groove side, to yield roll = 0°. Such effects have been seen for a few purine–pyrimidine steps in X-ray structures of DNA (see

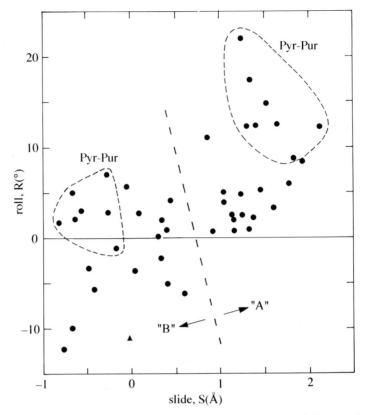

Figure 3.12 Plot of roll versus slide for many base-pair steps of oligomeric DNA as studied by X-ray diffraction up to about 1984. Points for pyrimidine–purine (Pyr–Pur) steps cluster in two discrete groups that correspond to the stacking geometries shown in Figs 3.10 and 3.11. ▲ is a Pyr–Pur step near one end of an oligomer which does not fit the general pattern. Data from recent structures (not shown) fit into the same broad pattern, except for a few Pyr–Pur steps with low propeller twist, and hence low roll at high slide. From Calladine and Drew (1984) *Journal of Molecular Biology* **178**, 773–82.

the few points in Fig. 3.12 near the lower left-hand corner). Presently, however, we think that the purine–pyrimidine step is not as important as the pyrimidine–purine step in determining the structure of DNA, because accurate, atomic models show that overlap of purines from opposite strands is not as great for a purine–pyrimidine step as for a pyrimidine–purine step, on account of the constraints provided by the sugar–phosphate chains.

Finally, could either roll or slide be coupled with *twist* in some way? From first principles, we might expect to see a relationship between slide and twist if the sugar–phosphate chains were semi-rigid and of constant length: the relevant geometry is shown in Fig. 3.13. In this example, the twist angle decreases from 36° to 28° as the base-pairs slide from 0 to +2 Å. There is a direct mechanical coupling involved here, because the sugar–phosphate chains are assumed to be rigid links of constant length. Actually, a plot analogous to the one shown in Fig. 3.12, but with twist instead of roll plotted along the vertical axis, does show a broad but definite tendency of the sort expected: low slide goes with high twist, while high slide goes with low twist. But the correlation of slide with twist is not so strong as the correlation of slide with roll, because sugar–phosphate chains are not actually as rigid as the picture of Fig. 3.13 suggests. In any case, it seems that roll, slide and twist in DNA are all related to one another.

Thus, having started with Euler's six degrees of freedom, and having eliminated three of these by introducing constraints on the base-stacking due to various factors, we find finally that the three remaining parameters R, S, and T are broadly related to each other in ways which depend on the base-composition of the step.

Now we have explained almost everything that is known today about the internal structure of DNA. In summary, the

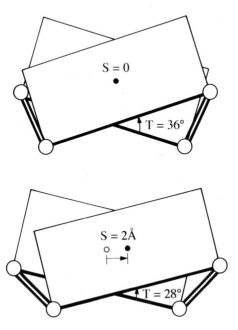

Figure 3.13 Schematic model of a base-pair step (cf. Fig. 2.5(b)) showing a possible mechanical linkage between slide and twist, which is confirmed weakly by experiment.

base-pairs adopt propeller twist to minimize their contact with water; this propeller twist prevents the otherwise flat base-pairs from sliding freely on one another's surfaces. Some base-pairs must roll apart from their neighbors at certain values of slide, depending on the base sequence. There are also some relations between slide and twist, that come from the connection of base-pairs to sugar–phosphate chains of roughly constant length.

Having completed our study of the patterns of base-stacking in DNA, our final task is to explain how different values of roll R, slide S and twist T *generate different kinds of double helix.* In other words, when the same values of roll, slide, and twist are repeated over and over again along a significant length of DNA, what kind of double helix will be formed? Once we understand this relationship between the "internal" variables R, S, T, which describe the base-stacking relationships, and the "external" form of the resulting helical structure, we shall be able to understand how different sequences of bases in DNA can generate different double-helical structures, by favoring different values of R, S, and T at a local level.

Our analysis will consist solely of three-dimensional geometry. For the present we shall assume that every step in a given helix has the same values of R, S, and T. An obvious way of proceeding might be to build some physical models of DNA, step-by-step, by means of a suitable home-made construction kit. That is exactly what we did ourselves in the first instance, when we were struggling to understand the geometry of double helices. From careful study of these models we were able to derive the relevant equations that describe the geometrical form of the DNA in three dimensions, as functions of R, S, and T. Do not worry if you cannot follow the details of our presentation: many people have difficulty with the simpler two-dimensional geometry that can be drawn on a piece of paper! The crucial point to grasp is that the final results could be established firmly by means of a few hours of practical construction at a woodwork bench.

The various stages of our analysis correspond to the pictures shown in Fig. 3.14(a)–(d). Let us look first at (a), which shows the side-view of a stack of 11 base-pairs, with $R = 0°$, $S = 0$ Å, and $T = 36°$ at each of the 10 steps. When roll and slide are both zero, then the helical geometry is very simple: neighboring base-pairs remain exactly parallel to each other, and exactly 10 steps are needed (with $T = 36°$) to complete one helical turn of $360°$. Hence, the side-views of the top and bottom blocks in Fig. 3.14(a) are the same. For these two particular blocks,

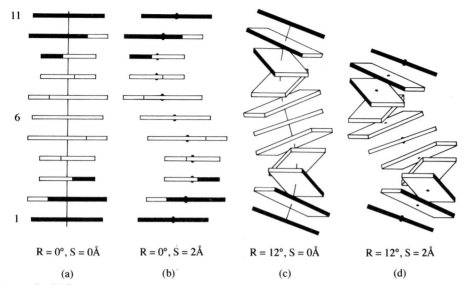

11

6

1

$R = 0°, S = 0Å$ $R = 0°, \dot{S} = 2Å$ $R = 12°, S = 0Å$ $R = 12°, S = 2Å$

(a)   (b)   (c)   (d)

Figure 3.14 One complete helical turn of DNA having $T = 36°$, showing the effects of introducing uniform roll R or slide S at each step. Broadly, (a) corresponds to the "B" form of DNA, while (d) corresponds to the "A" form as shown in Fig. 2.7. Parts (b) and (c) correspond to structures intermediate between "B" and "A" which have, in fact, been seen recently in DNA crystals by X-ray diffraction.

numbered 1 and 11, we get a full-frontal view of the minor-groove edge, which is here colored black. If we go along five steps from either end to base-pair 6, we find a block which is seen with a full-frontal view of the white, major-groove edge; in-between you should be able to make out the different edges of the blocks as they rotate about the axis.

Next, let us start from the simple helix shown in Fig. 3.14(a), having $R = 0°$, $S = 0$ Å, $T = 36°$, and then introduce a slide of $S = 2$ Å at each step. The resulting helical geometry is shown in Fig. 3.14(b), where base-pairs spiral outwards from a central helix axis. To understand this motion, we must study a top view of the structure as shown in Fig. 3.15. This diagram shows a half-turn of a helix in which the center of every block has moved outwards, radially from the axis by a distance r. By looking at the sideways displacement of the centers of the blocks, we can see that the outward motion is, in fact, associated with a slide S at each step. Applying trigonometry to Fig. 3.15, we find that

$$r = (S/2)/\sin(T/2).$$

Thus, given $S = 2$Å while $T = 36°$, the base-pairs move outwards by

$$r = (2/2)/\sin(36°/2) = 3.2 \text{ Å}$$

In summary, Fig. 3.15 shows that slide S makes a big hole in the middle of the helix, while the side-view of Fig. 3.14(b) shows that the stack of bases becomes wider than it was before.

What happens if the base-pairs roll apart from each other by some angle R, while $S = 0$ Å, and $T = 36°$? How will the picture of Fig. 3.14(a) change on account of the introduction of roll R? The resulting helical geometry is shown in Fig. 3.14(c), where each base-pair tilts from the vertical relative to its position in diagram (a). (In fact, each block rotates slightly about its own "front–back" axis.) This kind of motion may well be a surprise to you, but if you look carefully at (c) you will see that, at every step, the gap between the minor-groove edges is larger than that between the major-groove edges, corresponding to the positive roll of $R = 12°$ which has been used for the drawing. The cumulative effect of many roll angles is to cause a *tilt* with respect to the vertical axis. When the roll angles are small, it can be shown that

$$\text{tilt} \approx (R/2)/\sin(T/2)$$

Thus, when $R = 0°$, tilt $= 0°$; but when $R = 12°$ and $T = 36°$ then the tilt becomes $(12°/2)/\sin(36°/2) = 19°$, as shown in (c). Also, roll and tilt cause the base-pairs to move out slightly from the axis. The best way to understand this is to think about the

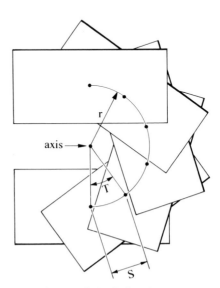

Figure 3.15 A top view of part of the helix shown in Fig. 3.14(b), illustrating the geometry by which radial displacement r is related to slide S and twist T.

thin black "rod" which connects the base-pairs in (a) and (c). In (a) it is straight, but in (c) the roll at each step curves the rod, and since the direction of curvature is different for each step, the rod ends up as a gentle spiral. Calculations show that the outward movement of base-pairs along the rod is approximately 3.3 sin (tilt)/2 sin $(T/2)$ = 1.7 Å.

So far we have considered combinations of just two parameters, S and T, to give distance from the axis as in (b), or R and T to give both distance from the axis and also tilt from the axis as in (c). But when all three parameters R, S, and T act simultaneously, something new happens: the helix gets shorter by S sin (tilt), because the sliding motion goes "downhill" on account of tilt. The relevant geometry is drawn in Fig. 3.16, and its effect on the overall structure may be seen in Fig. 3.14(d). The vertical stacking distance remains constant at 3.3 Å measured perpendicular to the blocks, but the rise h along the helix axis from the center of one base-pair to the next is reduced to

$$h = 3.3 \cos (\text{tilt}) - S \sin (\text{tilt})$$

With R = 12°, we know that tilt = 19°; and so with S = 2 Å we have

$$h = 3.3 \cos (19°) - 2 \sin (19°)$$
$$= 3.1 - 0.6 = 2.5 \text{ Å}$$

To summarize, we can calculate approximately the overall shape of a long helix made out of many uniform (R, S, T) steps according to three formulas:

(i) distance from axis =
   $(S/2) \cos (\text{tilt})/\sin (T/2) + 3.3 \sin (\text{tilt})/2 \sin (T/2)$,
(ii) tilt from axis = $(R/2)/\sin (T/2)$,
(iii) length along axis = $3.3 \cos (\text{tilt}) - S \sin (\text{tilt})$.

We are now ready to relate different values of R, S, and T, as determined by the base sequence of DNA, to the overall shape of a double helix. For example, in Fig. 3.6 we showed that AA/TT prefers to stack so that it can form an extra hydrogen bond in the major groove; values of (R, S, T) for an AA/TT step in its preferred arrangement are (0°, 0 Å, 36°). Therefore, a double helix containing only AA/TT steps will resemble Fig. 3.14(a), except, of course, that the base-pairs will be highly propeller-twisted: remember that the drawings shown in Fig. 3.14 do not show the propeller twist. As another example, the GG/CC step seems to prefer high slide, S = +1 to +2 Å, because of a strong imbalance of electric charge within the

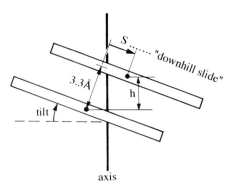

Figure 3.16 For base-pairs at a standard perpendicular separation of 3.3 Å, the "rise" *h* in the direction of the helix axis depends both on tilt and slide *S*.

guanine ring that discourages one guanine from stacking directly over another (see Appendix 2, Fig. A3). Therefore, a helix containing all GG/CC steps will often resemble Fig. 3.14(b) or (d), both of which show high-slide arrangements. The roll angle for GG/CC is actually around +6°, or halfway between the roll angles of 0° and +12° which were used to construct diagrams (b) and (d), so the GG/CC helix should lie roughly halfway in structure between (b) and (d).[4]

As a final example, recall that pyrimidine–purine steps prefer either of two different stacking arrangements, shown in Figs 3.10 and 3.11, rather than a continuous range. Roll–slide values for these two preferred arrangements are $(R, S) = (0°, -1 \text{ Å})$ and $(+20°, +2\text{Å})$. One can imagine that a double helix containing a substantial fraction of pyrimidine–purine steps might be forced or aligned into either of two shapes by the "bistable" nature of stacking at the pyrimidine–purine steps. Indeed, such helices do have two separate shapes known as "B" and "A": see Fig. 2.7. The "B" form exhibits $(R, S) = (0°, 0 \text{ Å})$ as an average over many steps, while the "A" form shows $(R, S) = (+12°, +1.5 \text{ Å})$. We think that the bistability of pyrimidine–purine steps on a local level may be responsible, at least in part, for the distinct "B" and "A" shapes of DNA on a large scale. Almost all of the external features of "B" and "A" helices, such as the distance of pairs from an axis, the tilt of pairs with respect to an axis, and the rise along the axis, can be calculated from their (R, S, T) values according to the formulas given above.

These few examples of the use of (R, S, T) values in understanding DNA have related mainly to physical measurements of DNA structure from X-ray studies, rather than to the role of DNA in biology. But we must start somewhere! In the following chapters we explain how the roll–slide–twist model is indispensable for understanding many of the roles of DNA in biology, such as how promoters work, how DNA coils around proteins in a chromosome, and how DNA binds gene-regulatory proteins such as "repressors" or "activators." Don't worry too much about the details of this chapter, such as the various formulas and constructions, so long as you learn the meanings of propeller twist, and of roll, slide and twist.

You may be puzzled that we have not shown the sugar–phosphate chains in any of the pictures from this chapter, except in Fig. 3.13. The chains are there, of course, but they have not been shown in the drawings. This is analogous to the way in which we have not shown propeller twist in some of the diagrams, either. The really important point in the present chapter is that the outward features of DNA *all* depend strongly on base-stacking arrangements at the inner core of the molecule. In Appendix 2 we describe more fully the chemical forces which influence how the bases stack on one another, for the benefit of those of our readers who are trained in chemistry or physics.

Notes

1. See Appendix 1.
2. In this chapter, we have used the symbols R, S, and T for roll, slide, and twist, respectively, with the sign conventions shown in Fig. 3.8. In fact, many different symbols and sign conventions have been used for these quantities by different scientists, and we have chosen the present set for the sake of simplicity in a textbook. Our symbols can be translated into those of the X-ray diffraction workers (Dickerson *et al.*, 1989) as follows: $R = \rho$, $S = -D_y$, and $T = \omega$.
3. See Appendix 2 for a more complete explanation of the physical forces that influence base-stacking geometries in DNA. It seems likely that partial electrical charges within the base-pairs, as shown in Fig. A3, may contribute substantially to the bistable stacking behavior of pyrimidine–purine steps (C.A. Hunter, personal communication).
4. Recent work by Dietrich Suck shows that these high-slide structures persist in complexes of DNA with the protein DNAase I.

Further Reading

Calladine, C.R. (1982) Mechanics of sequence-dependent stacking of bases in B–DNA. *Journal of Molecular Biology* **161**, 343–52. Steric consequences of propeller twist for the stacking of base-pairs in the "B" form of DNA.

Calladine, C.R. and Drew, H.R. (1984) A base-centred explanation of the B-to-A transition in DNA. *Journal of Molecular Biology* **178**, 773–82. First full statement of the roll–slide–twist model for DNA, as generalized to all right-handed forms of the molecule.

Dickerson, R.E. *et al.* (1989) Definitions and nomenclature of nucleic acid structure parameters. *EMBO Journal* **8**, 1–4. A comprehensive listing of possible structural parameters for DNA. Note that propeller twist and slide are reversed in sign here as compared to many papers in the literature, and all figures in this book.

Hogan, M., Dattagupta, N., and Crothers, D.M. (1978) Transient electric dichroism of rod-like DNA molecules. *Proceedings of the National Academy of Sciences, USA* **75**, 195–99. Early solution data favoring the existence of propeller twist in DNA base-pairs.

Levitt, M. (1978) How many base-pairs per turn does DNA have in solution and in chromatin? Some theoretical calculations. *Proceedings of the National Academy of Sciences, USA* **75**, 640–44. Early theoretical calculations favoring propeller twist in DNA.

McCall, M.J., Brown, T., and Kennard, O. (1985) The crystal structure of d(GGGGCCCC): a model for poly(dG).poly(dC). *Journal of Molecular Biology* **183**, 385–96. A summary of observations concerning high-slide at GG/CC steps, as seen directly by X-ray diffraction.

Nelson, H.C.M., Finch, J.T., Luisi, B.F., and Klug, A. (1987) The structure of an oligo (dA).oligo(dT) tract and its biological implications. *Nature* **330**, 221–6. Direct observation of very-high propeller twist in a series of A–T base-pairs, and a postulate of an additional hydrogen bond in the major groove between adjoining pairs.

Wing, R.M., Drew, H.R., Takano, T., Broka, C., Tanaka, S., Itakura, K., and Dickerson, R.E. (1980) Crystal structure analysis of a complete turn of B–DNA. *Nature* **287**, 755–8. First direct structural evidence for propeller twist in DNA base-pairs, by X-ray diffraction.

Exercises

3.1 Normally the DNA double helix is right-handed, as shown schematically in Fig. 3.1. In this case, the provision of left-handed or counterclockwise propeller twist (when looking along the long axis of any base-pair), as shown in Fig. 3.4, can reduce the access of water to the bases, as shown in Fig. 3.3.

For a hypothetical *left-handed* double helix of DNA, with

$T = -32°$, what sense of propeller twist would be required to reduce likewise the access of water to the bases?

(Note: the left-handed "Z" DNA shown in Fig. 2.7 has almost no propeller twist, because the bases there stack not only onto neighboring bases, but also onto neighboring sugars.)

3.2 In the models of the "A" and "B" forms of DNA shown in Fig. 2.7, the *major*-groove edges of the bases are shaded heavily— a convention which is opposite from that used elsewhere in this book. Identify the major and minor grooves which lie between the sugar–phosphate chains in these two models. In the "B" form, which groove has the larger width? In the "A" form, which groove is wider? Or are the widths about the same? In the "A" form, which groove is *deeper*?

3.3 Using Fig. 3.5 as a guide, identify the major- and minor-groove edges of the base-pairs shown with atomic detail in Fig. 2.11(a) and (b).

3.4 L. Euler explained long ago that any one rigid block has 6 degrees of freedom of motion with respect to another rigid block. Each of these may be described in terms of a translation along, or a rotation about, any of the three axes which are labeled in Fig. 3.7. In practice, 3 of these 6 degrees of freedom are *not* mobilized significantly in the base-pair steps of DNA.

By use of a simple model involving blocks of wood, or cardboard boxes, confirm that translation or shortening along the "twist axis," and rotation about the "front–back" axis, are inhibited because of the close surface-to-surface stacking of the bases.

Also, examine the third unused degree of freedom, which involves translation along the front–back axis. It is not clear whether this motion is inhibited in real DNA by the chemical forces which influence base stacking, or by the action of sugar–phosphate chains, or both. The front–back motion could conceivably be favored for certain sequences in DNA, but we have at present no good examples.

Finally, confirm that the three allowed motions of roll, slide, and twist in DNA, as shown in Fig. 3.8, are also allowed in the wooden-block models.

3.5 a   Make a simple physical model to illustrate the linkage between slide and twist shown in Fig. 3.13. (Expanded polystyrene foam can be cut easily into suitable blocks using a bread-knife, and the sugar–phosphate chain links may be made from wires, with their ends pushed into the blocks.)

b Make a simple physical model of two adjacent propeller-twisted base-pairs, with the bases of unequal size as in Fig. 3.6. (Expanded polystyrene foam blocks may be held apart conveniently, in a propeller-twisted arrangement, by means of cocktail sticks.) Use these model base-pairs, without any sugar–phosphate chains, to study the linked slide–roll motion which is shown in Figs 3.10 and 3.11.

3.6 The table below gives approximate, uniform values for roll, slide, and twist at the base-pair steps in three different well-known forms of DNA—"A," "B," and "C"—which have been studied by X-ray diffraction of fibers:

	"A"	"B"	"C"
R (°)	+12	0	−6
S (Å)	+1.5	0	−1
T (°)	32	36	40

Use the equations on p. 59 to calculate approximate values of the following parameters for each of the three forms "A," "B," and "C":

a the distance of the centers of base-pairs from the axis of the double helix;

b the angle by which base-pairs are tilted from planes normal to this axis; and

c the length, or "rise," of the molecule per base-pair along the axis.

(Note: calculate the tilt angle first.)

CHAPTER 4

Twisting and Curving

In the last two chapters we have learned some rudimentary things about DNA. We have learned: (a) why DNA forms a helix, (b) how the bases make ordered pairs at the center of the helix, (c) how the bases twist like a propeller within any base-pair, and (d) how the overall shape of a helix depends on the local parameters roll, slide, and twist over a series of base-pair steps. This is the stuff of chemistry, not biology. When are we going to start talking about DNA in biology? That is exactly what we shall be doing in this chapter.

The two most fundamental actions of DNA in biology involve either the *twisting* or *curving* of a DNA double helix. First, we consider the twisting of DNA—or, to be more precise, its *un*twisting. There are two main instances where DNA has to untwist as it carries out its duties in a cell: first, when DNA is copied into the messenger RNA that tells the cell how to make protein; and second, when DNA is copied into another DNA strand just before a single cell divides into two cells. The first of these processes is called "transcription," and the second is called "replication": see Fig. 4.1. In each case, the DNA unwinds into two separate sugar–phosphate strands. These pictures provide, of course, only a static representation of a brief instant in the life of a cell. You have to imagine that the unwound regions of DNA in pictures (a) and (b) are moving rapidly across the page, from right to left, in order to grasp the dynamic nature of DNA unwinding in living systems.

Often DNA unwinds only over a short region, say 15–20 base-pairs when making RNA as in (a), because it takes a lot of energy to pull the base-pairs apart and expose them to water. The "bubble" of unpaired bases can travel along the length of the DNA very rapidly, at about 100 base-pairs per second; but then, time doesn't mean much to these tiny molecules that we can hardly see by using a light microscope. When DNA gets copied into RNA, a copying protein or enzyme called "RNA

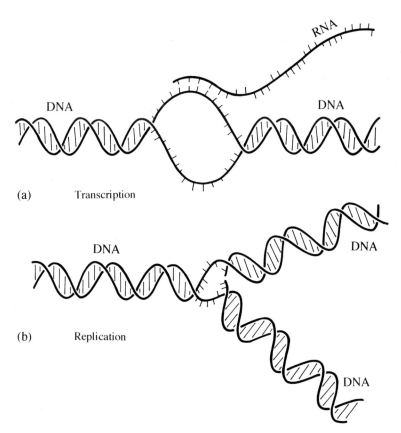

Figure 4.1 Schematic representations of transcription and replication of DNA. In each case, the DNA must untwist locally to let one strand serve as a template for the synthesis of a new strand, either of (a) RNA or (b) DNA.

polymerase" attaches itself to one of the two DNA strands: see Fig. 4.2. Then, in a way that no one fully understands, the polymerase pulls nucleotides out of solution to match the bases which it finds within the DNA chain. In Fig. 4.2 the enzyme is just about to add an RNA base C to a DNA base G. Another somewhat similar enzyme carries out the process of copying DNA into DNA (Fig. 4.1(b)); it is called "DNA polymerase." Any cell contains several varieties of RNA polymerase, and several varieties of DNA polymerase, to do different kinds of copying tasks.

These RNA polymerases in the cell always try to put C with G, G with C, A with T, and U with A to make Watson–Crick base-pairs, as shown inside the "bubble" of Fig. 4.2. The DNA polymerases do the same, but add T instead of U to A. Some of the copying enzymes, in a test-tube, will incorrectly put T

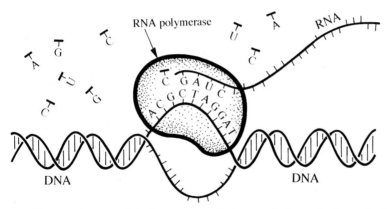

Figure 4.2 An RNA strand in the process of being made from single nucleotides by the enzyme RNA polymerase. One of the two DNA strands serves as a template for accurate synthesis of the new RNA strand, by the rules of Watson–Crick pairing.

with G if you feed them a lot of T nucleotides and no C at all; but they add T to G much more slowly than C to G. These kinds of copying error do not happen very often in living cells, where all four kinds of nucleotide are relatively abundant.

Now, you remember that DNA is helical, like a screw: so, as the RNA polymerase and its "bubble" of unpaired bases move along the DNA, either the polymerase must screw *around* the DNA, or the DNA thread must screw itself *through* a stationary polymerase. Which of these alternatives will be favored in living cells? The two competing models are shown in Fig. 4.3, and it is not obvious on first inspection which might be right. Just recently, James Wang and Leroy Liu have shown that the polymerase remains somewhat stationary, as in Fig. 4.3(b), while the DNA screws through it. There are good reasons why this should be so. First, it will be hard for the big polymerase protein to move rapidly through the sticky, viscous fluid of a cell nucleus, in order to rotate about the DNA by 10 times per second; the thin, wiry DNA can rotate in the fluid much more easily. Second, the polymerase drags behind it a long RNA "tail" as it goes about its duties, as shown in Fig. 4.2. It would be very hard for this long tail of perhaps 500–1000 nucleotides to follow the polymerase round and round the DNA, at a rate of 10 times per second. Indeed, some experiments suggest that in a test-tube the polymerase finds it easier to rotate about the DNA if the RNA tail is cut off by another enzyme (called RNAase) as it is being made.

If you have ever been out fishing, and the current twists your

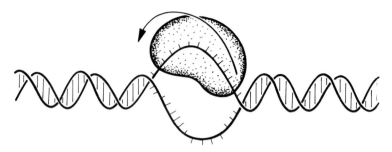

(a) DNA fixed, polymerase rotates

(b) Polymerase fixed, DNA rotates

Figure 4.3 Alternative schemes for the copying of DNA into RNA by the enzyme RNA polymerase. In (a), the polymerase screws around a stationary DNA, while in (b) the DNA screws through a stationary polymerase.

bait around the end of your line, you will know that there will be an unholy mess. Similarly, it would cause big trouble if Nature did not find some way to relieve the twisting stress of DNA during its passage through a polymerase. In fact, Nature has invented several different kinds of de-knotting enzyme, called "topoisomerases,"[1] to get rid of the excess DNA twist. These enzymes act in very subtle ways, and no one is quite sure how they work. They cut either one or both of the two sugar–phosphate chains to allow some kind of motion in the DNA to relieve torsional stress; then they re-connect the broken parts and thus leave an intact but relaxed DNA. Repeated application of this process to a tangled-up piece of DNA will eventually unravel it. We discuss the tangling (or supercoiling) of DNA further in Chapter 6.

Some anti-cancer drugs are directed against the de-knotting enzymes. The cancer cells, which are growing out of control, can't divide if you poison their de-knotting enzymes. Many anti-cancer drugs "trap" the topoisomerases in the middle of their cutting–relaxing–rejoining cycle, so that they cannot rejoin the strands of the DNA. Apparently, those cancer cells which survive the first few rounds of treatment with these drugs somehow find a way to grow with only a very low level of de-knotting enzymes; and so they are eventually able to resist further treatment.

It certainly doesn't look very promising, if you are a biological scientist today, to look for more anti-cancer drugs that poison topoisomerases. In fact, the only really successful anti-cancer drug discovered in the past 20 years was found by Barnett Rosenberg, who was studying how bacteria grow in an electric field. He found that certain electrodes cause bacteria to grow long and thin, like spaghetti. By lots of detective work, he eventually worked out that the platinum in his electrode was combining with ammonia in his buffer, to make a platinum–ammonia compound that prevented cell division. This compound had been sitting on people's shelves for over 100 years, but no one had ever thought it would cure cancer. Anyway, he tested it on people, and it worked in some cases. Apparently, the platinum atom uses two of its ligand-binding sites to cross-link the two DNA strands, so that the separation of strands which is needed for replication (see Fig. 4.1(b)) cannot occur.

We mentioned above that DNA unwinds into a bubble of 15–20 bases, as RNA polymerase copies one strand of the DNA to RNA. How is this accomplished? It turns out that the hard step is to unwind the double helix to begin with, in just one location. After that has been done, the bubble of DNA can travel with relative ease to some other location along the length of the molecule. It is therefore of crucial importance to understand how the cell tells its DNA to unwind in certain, specific locations. These events are then responsible for the large-scale synthesis of RNA from many particular genes on the DNA; and so they are responsible, ultimately, for the kinds of protein which are found in any cell. A similar kind of unwinding takes place at the start-sites for replication, where both strands of the DNA are copied into DNA; only there the enzymes are different.

What would you do if you were faced with the problem of unwinding the DNA from a cell in many specific locations? Would you make a protein with a little shovel attached at one end, to dig or pry a hole in the DNA? Probably some DNA-unwinding proteins do have little "shovels," in the form of flat, oily amino acids such as phenylalanine, tyrosine, and tryptophan, which can make holes and then insert themselves between the base-pairs, and thereby effect unwinding of the helix. They would act just like the ethidium bromide molecule shown in Fig. 2.9, which converts the DNA locally from a helix into a partially untwisted ladder. Other DNA-unwinding proteins perhaps have clefts on their surfaces into which only one of the two DNA strands can fit, so that they bind to a single strand in preference to a double helix, and so unwind the molecule that way. Finally, yet other proteins might curve the DNA around themselves in the form of a telephone cord or a coiled bed-spring, so that the DNA can unwind as it vibrates like a concertina in solution. There is some evidence for all of these kinds of unwinding in the cell, but at present there is little knowledge at a detailed, atomic level of what happens in any case.

All of these unwinding events involve the binding of some protein to DNA. More is known about the role of DNA in such interactions than about the role of the protein. As a general rule, Nature makes the double-helical connections very weak in places where unwinding must begin, so that it won't take much energy to break open the DNA base-pairs, and thereby separate the two strands from one another. You will recall from Chapters 2 and 3 that the main determinants of the stability of a double helix are the number of hydrogen bonds between the bases in a base-pair, and the extent of base-to-base overlap. An adenine–thymine pair has only two hydrogen bonds, while a guanine–cytosine pair has three (see Fig. 2.11); so a series of adenine–thymine pairs will make the double helix less stable, as shown in Fig. 4.4(a). Furthermore, it has been shown by many careful experiments that pyrimidine–purine sequences have the least stability with regard to base-to-base overlap. These are the steps TA, TG, CA and CG. Why should this be so?

In the last chapter, we explained how to describe DNA in terms of the roll R, slide S and twist T at any base-pair step. Pyrimidine–purine steps are special in this description because they can adopt either of two stackings: a low-slide form where the purines slide apart from one another (Fig. 3.10), or a high-slide form where the purines slide on top of one another (Fig. 3.11). Now the base-pairs are connected to sugar–phosphate chains in such a way that *high slide* leads to *low twist*. Thus, in the example shown in Fig. 3.13, a slide of $S = +2$ Å leads to a low twist of $T = 28°$ or thereabouts, as compared to $T = 34–36°$ for other DNA. So it may be that the amount of energy required to unwind the DNA further, say to $T = 10°$ or $20°$, will be less for a pyrimidine–purine sequence than for other DNA, because the base-pairs there are already unwound to a significant degree.

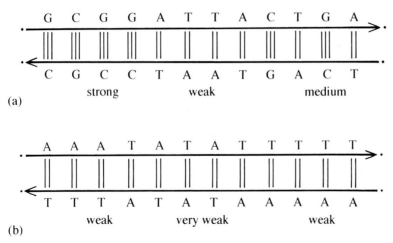

Figure 4.4 A two-part explanation for the ease of DNA unwinding at any TATA sequence. In (a), the double or triple cross-chain lines represent 2 or 3 Watson–Crick hydrogen bonds, respectively; and the double helix is less stable with 2 bonds than with 3. In (b), the TA steps (of kind pyrimidine–purine) cause a further weakening of the cross-chain connections, because they can easily unwind to yield low twist.

In any case, there seems to be little energetic barrier to low twist at a pyrimidine–purine step, for reasons that may not yet be fully understood.

Which, then, are the most easily unwound sequences in DNA? Simply those which combine the two characteristics which we have been describing: few hydrogen bonds as in A–T pairs, and low twist as in pyrimidine–purine steps. In other words, DNA unwinds most easily at AT-rich regions that have many pyrimidine–purine steps, as shown in Fig. 4.4(b). The prototypic "weak" sequence is something like "TATATATA" or "TAATAA-TAA," where TA is the pyrimidine–purine step of low twist.

This is not just a theoretical, hand-waving argument: there is actually strong experimental evidence for a low-twist intermediate in the unwinding of TATA-type sequences. It seems that such sequences unwind to "cruciform" configurations more easily than other DNA. Now a cruciform is just a big "bubble" of DNA which can be trapped and studied, because each strand of the bubble folds back and pairs with itself to form a double helix, as shown in Fig. 4.5. James McClellan and David Lilley set out to determine the reason for this odd behavior. They found that the unwinding of DNA into a cruciform at TATA-like sequences was being catalyzed by partial unwinding of the DNA from 10.5 to about 12 base-pairs per turn, or from $T = 34°$ to about $T = 30°$, before the cruciform appeared. Their observations agree closely with what we expect from theory, as explained above.

Biologists identified the TATA-type sequences almost by accident, by determining the sequences of DNA in places where the double helix must unwind for transcription to begin. These broadly defined regions of DNA are called "promoters," because they promote RNA synthesis from a nearby gene without being transcribed themselves. The biologists found in many cases a "TATA" or similar sequence at the required site of unwinding within each promoter. Then they drew a box around this sequence on their computer output, to show how important it might be. Hence, many textbooks call the weak region of DNA a "TATA box." The majority of promoters contain a TATA or related sequence that helps the DNA to unwind.

In most bacteria, an RNA polymerase protein seems to bind to the TATA region in front of a gene directly, without any intermediary protein to help unwind the TATA sequence for it. But in the cells of our body, there is an additional protein called "TFIID" (for "transcription factor, polymerase II, fraction D") which binds to TATA sequences near the start of genes. This protein is available in moderate amounts in pure form, but no one has yet determined whether it unwinds the TATA sequence in preparation for the polymerase, as one might suspect.[2]

There is strong evidence for the role of a TATA sequence in DNA unwinding, from studies of gene function in bacteria. There the initial sites of unwinding are known in many cases,

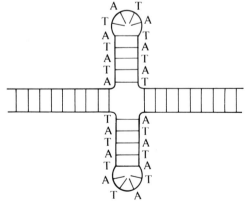

Figure 4.5 A "cruciform" structure which has been "extruded" from a suitable sequence of A and T bases. The DNA double helix unwinds from 10.5 to 12 base-pairs per helical turn before the cruciform appears, owing to the easy unwinding of TATA sequences. As a consequence of this initial unwinding, the cruciform extrudes there more readily than at other sequences.

and they usually correspond to TATA-like sequences. But the situation in higher organisms is not yet clear, because there the initial sites of DNA unwinding to start transcription are not known. Some genes in higher organisms contain a prominent TATA-like sequence near the startpoint for making RNA, whereas others do not. We expect, in any case, once the initial sites of unwinding have been determined, that they will correspond at least to moderately weak double-helical structures such as TATG, TAAA, or even TGTG, because it will be hard for a protein to unwind the DNA if this is not the case.

Thus far we have assumed that a TATA sequence must unwind, perhaps in a stepwise fashion, from a double helix to a single strand as if no other choice were available. But actually a TATA sequence can adopt another, entirely different helical form which has not yet been well-characterized: one with 7.5–8 base-pairs per turn rather than 10–12 for normal DNA. This helix has been called the "D" or "E" form of DNA in X-ray studies of DNA fibers, at low resolution. Until we establish the molecular structure of such a strange helix, we cannot even guess whether it might be used in biology.

As a last note on the subject of the weakness of TATA-type double helices, we refer to the three triplets in the Genetic Code (Table 1.1) which code for "stop," and are usually called "stop-codons." The process of assembling amino acids into a protein chain comes to a halt because there are no transfer RNA molecules which can recognize these particular triplets. One of these codons is UAA in the messenger RNA (like TAA in DNA), and if a transfer RNA molecule were to exist, it would form a specially weak bond with this triplet (see Fig. 4.6) for precisely the reasons given above. We think that it is no accident that this specially weak-binding triplet does not correspond to a specific amino acid, for the hypothetical transfer RNA would be unreliable at recognizing this triplet. We think that the same considerations probably apply to the other two stop-codons UAG and UGA, which would also be moderately weak in their capacity for base-pairing.

That gives an overview of twisting in DNA: the local unwinding of DNA to form a "bubble"; the passage of DNA like a screw through a polymerase; and the weak TATA sequence that helps the bubble to form. Now let us look at the curving of DNA. In principle, DNA could curve either in a plane, like a banana, or else in three dimensions, like a coiled bed-spring. It is hard to learn about three-dimensional curvature at the first attempt, however; so we restrict ourselves in this chapter to the

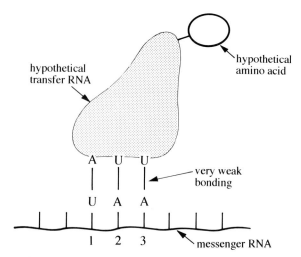

Figure 4.6 All of the "STOP" signals in the Genetic Code (UAA, UAG, UGA, see Table 1.1) would correspond to very weak or moderately weak pairings between a hypothetical transfer RNA and the messenger RNA.

curving of DNA in a plane. The next chapter will deal with curving in three dimensions.

First let us forget about DNA, and think instead for a moment about our gardens. Many people have small, flowering trees in their suburban gardens, and they like to use slabs to build neat paths around them, as shown in Fig. 4.7. There are two ways to do this. One way is to buy special slabs that are made in a wedge-shape, with one end wider than the other. Such slabs will make a circle of a particular radius, just as stone "voussoirs" will build a round arch of specific radius over a window or a door. The other way is to make the path out of rectangular slabs, and to fill the narrow triangular spaces between them with pebbles. An advantage of this scheme, of course, is that one can change the curvature of the path at will, making it circular, or straight, or of variable curvature as the plan requires, just as in the picture.

What do we mean by the *curvature* of a garden path? In Fig. 4.7 the path is obviously more curved where the angle between consecutive slabs is larger. Where this angle is zero, the path is straight; that is, it has zero curvature. It would therefore be sensible to define curvature in terms of the angle between successive slabs. Thus the curvature of the semicircular part of the path in Fig. 4.7 may be described as 15° per slab (or per step) because the path turns through 180° in 12 steps.

Now, if we lay a garden path with 15° between every pair of slabs, we shall eventually complete a circle. And the radius of

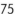

Figure 4.7 Curvature of a garden path in a plane, around a tree.

the circle will be smaller if we make the angle of curvature larger, and vice versa. We can easily calculate the radius of the semicircle in Fig. 4.7, as follows. Suppose the width of each slab is 0.5 m. If the angle between slabs is 15°, there will be 360/15 = 24 slabs in a circle, with a total circumference of 24 × 0.5 = 12 m, and hence a radius of 12/2π = 1.9 m. This is the radius of the inner edge of the slabs; similarly we could find the radius of the center-line of the path, by using the center-to-center separation of the slabs, together with the angle between slabs.

The curvature of DNA is more complicated than that of a garden path, because DNA is three-dimensional; yet it follows broadly the same principles, because DNA is also made in discrete steps, being built up from base-pairs. Imagine that a piece of DNA of length 80 base-pairs, or 8 double-helical turns, has been bent into a 360° circle. That is just about the degree of curvature by which DNA wraps around proteins in the cell nucleus (Fig. 1.5). If you were to make the same circle from eight tiny slabs, then the angle between slabs would have to be 360°/8 = 45°. This angle is precisely equivalent to the roll angle R in DNA, as described in Chapter 3. If we wish to make a circle from eight helical turns of DNA, we can do so by putting one roll angle of 45° in each double-helical turn: that will make a fine circle.

Such a scheme is shown in Fig. 4.8(a), which shows 20 base-pair steps that make up one-quarter of our 80-base-pair circle.

In order to convert straight DNA into curved DNA, we have introduced a roll angle of 45° at the two steps labeled 0 and 10, respectively, on the left-hand side of the drawing. These roll angles open up the minor-groove edges of the base-pairs, which are coloured black, to yield a total curvature of $2 \times 45° = 90°$. The same type of curvature is shown schematically in Fig. 4.9(a), which gives a plot of roll angle versus step-number in the sequence: roll angles of 45° can be seen at steps 0 and 10 in this diagram.

It would be better, however, to spread these roll angles over a greater number of base-pair steps within the circle. While a single roll angle of 45° per helix turn is satisfactory from a geometrical point of view, it is not satisfactory from a physical point of view, because the opening of a single base-pair step by 45°, as in Fig. 4.8(a), would expose a lot of water-insoluble base surfaces to the solvent. A slight improvement might be made by dividing the 45° into two parts, and thus locating a roll angle of 22.5° at two places within each double-helical turn. This scheme is shown in Fig. 4.8(b), and it has also been plotted in Fig. 4.9(b): roll angles of +22.5° are located at steps 0 and 10, while roll angles of −22.5° are located at steps −5, 5 and 15.

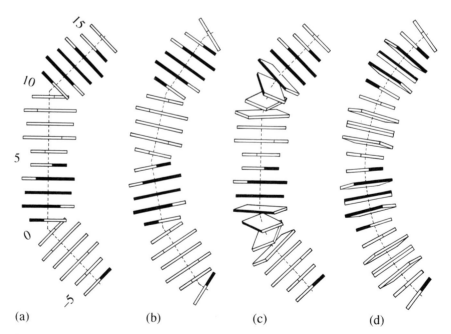

(a)   (b)   (c)   (d)

Figure 4.8 Two complete helical turns of DNA, with a curvature of 45° per turn, or 4.5° per step on average. Such tight curvature may be achieved, in principle, by any of the distributions of roll angle shown in parts (a) to (d).

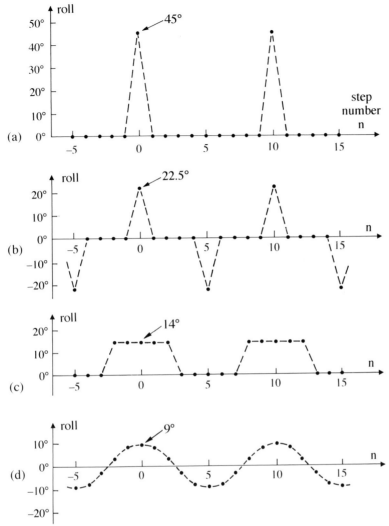

Figure 4.9 A plot of roll angle versus step number for the four cases shown in Fig. 4.8, each of which curves the DNA by 45° per helical turn.

The sign of the roll angle changes from plus to minus every five base-pairs, because the helix rotates by 180° over that distance. So the two angles of 22.5° open up the same side of the helix, and make equal contributions to the curvature despite the change in sign.

Another way of bending DNA into the required curve might be to have five steps with a roll of about 14° followed by five with zero roll, and then five more with $R = 14°$, etc. This new scheme is shown in Figs 4.8(c) and 4.9(c). In effect, we are converting successive half-turns of DNA into different configurations; and it is not difficult to see that this will produce a curve. The model shown in Fig. 4.8(c) is sometimes called a "junction" model for DNA curvature, because "bends" seem to appear at the junctions between the successive portions of DNA, each having uniform but different roll.

Perhaps the best scheme of all for making curved DNA would be to spread the roll over the entire helix, at every base-pair step, rather than placing big roll angles at only a few steps. This idealized model is shown in Fig. 4.8(d), and in the corresponding diagram of Fig. 4.9(d). If we let the roll angles vary as a cosine wave, starting at step 0 and ending at step 10, then we shall need to vary these roll angles by no more than plus or minus 9° to get a curvature of 45° per turn. You can see that there are much less abrupt changes in roll from step to step in Fig. 4.8(d) than in any of (a), (b), or (c). This may seem a bit like magic: how can we know what curvature will result from such an assortment of different roll angles as in this example?

The roll angles in Fig. 4.9(d) vary as a cosine wave of amplitude 9°, or as $R_n = 9°\cos(36°n)$, where $n = 0, 1, 2, \ldots,$ 9 identifies each step in any double-helical turn. For example, the roll at step 0 is $R_0 = 9°$, while the roll at step 1 is $R_1 = 9°\cos(36°) = 7.3°$, and the roll at step 5 is $R_5 = 9°\cos(180°) = -9°$. This new arrangement is similar overall to the many schemes shown before in (a), (b), and (c), but it is more subtle in the way in which it assigns roll to different parts of the helix. Now we come to a crucial point: not all roll angles contribute fully to curvature, because some of them cause the DNA to bend in the wrong way, out of the plane of the paper in Fig. 4.8. Let us look again at the helix shown in Fig. 4.8(a). In this picture, a roll angle at step 2, between steps 0 and 5, would bend the DNA down into the paper, rather than to the right as desired. But a roll angle at step 0 bends the DNA to the right, while a roll at step 5, while not shown in the drawing, would also bend the DNA to the left or to the right, as shown in Fig. 4.8(b). It can be shown that the contribution to rightward curvature by any step is its roll angle R_n, multiplied by the cosine of the total helix twist, or $R_n \cos(36°n)$. Thus, with arrangement (d) we get for step 0 a big contribution to curvature of $9°\cos(0°) = 9°$, and for step 5 we also get a big contribution of $-9°\cos(180°) = 9°$; but for step 2 we get only the small contribution of $9°\cos^2(72°) = 0.9°$, because the roll points in the wrong direction.

It follows, then, that the overall curvature k due to roll at all steps $n = 0, 1, 2, \ldots, 9$ can be found by adding together 10 different terms of the kind $R_n \cos(36°n)$, or in this case

$$9° \cos^2 (36°n) = 9° + 5.9° + 0.9° + 0.9° + 5.9° + 9° + 5.9°$$
$$+ 0.9° + 0.9° + 5.9° = 45°$$

In summary, the roll angles in Fig. 4.9(d) vary as $9° \cos \theta$, where θ is the total twist (or sum of base-step twists T) relative to step 0; and then you have to multiply them by $\cos \theta$ again before adding up to get the curvature, because not all steps point in the same direction. This simple scheme is not quite exact, geometrically, but it works well for small roll angles of less than or equal to about 10°, and that covers most practical cases.

It is not difficult to show that if a constant roll angle is added to every step, so that $R_n = 9° \cos (36°n) + $ (constant), then the sum worked out above has exactly the same value as before. A constant change of roll at all steps will change the overall appearance of the DNA, just as in Fig. 3.14(c), but it imparts no curvature.

As we have seen, not every step contributes equally to curvature in our scheme. Steps 0, 1, 4, 5, 6, 9 (where the minor or major groove faces the center of curvature) contribute a lot, while steps 2, 3, 7, 8 don't contribute much at all. The reason for this behavior is that the DNA is not flexible enough in a direction at right-angles to the roll axis (that is, about the "front–back" axis of Fig. 3.7) to curve very much in that direction. Also, one might suspect that slide S and twist T vary along with roll R as the DNA curves, in accord with the relations among R, S, and T discussed above; but this would make only small differences to the general pattern of behavior which we have described.

When studying the curvature of real DNA, which is made from a variety of base sequences, we cannot expect every kind of sequence to follow a smooth, wave-like pattern of roll angles exactly. Some steps, such as the pyrimidine–purine variety discussed in Chapter 3, are flexible enough to adopt either high roll or low roll with little difficulty. However, others such as AA/TT are known from studies of DNA in the crystal to prefer low roll, $R = 0°$; while steps such as GC/GC are thought to prefer high roll, $R = +5°$ to $+10°$. Essentially, what we must do for real DNA, when curved around a protein spool, is to find the best fit of a cosine wave to a given series of R values, at a period of 10 steps, even though some of these steps may not agree with the idealized cosine wave exactly. The spread of roll angles can range from $+9°$ to $-9°$, or from $+12°$ to $-6°$; it doesn't matter. Once we have obtained a "best fit," then the amplitude of this best-fit cosine wave will be proportional to the absolute curvature of the sequence, while its left-to-right position or phase, as in Fig. 4.9(d), will tell which steps have the largest (or most positive) roll angle, and which steps have the smallest (or most negative). Where the roll angle is large and positive, the minor-groove edges of base-pairs will lie along the *outside* of the curved DNA, and where the roll angle is small (or most negative), the minor-groove edges will lie along the *inside* of the curve. This protocol is known to mathematicians as taking the "Fourier transform" of the roll angles, to get the amplitude and phase of helix curvature.

You can see for yourself how this works by making a detailed plot along the lines of Fig. 4.9(c) for a series of steps such as those shown in Fig. 4.8(c). There, a batch of 5 steps (-2, -1, 0, 1, 2) with $R = 14°$ is followed by a batch of 5 steps with $R = 0°$, and then by another batch with $R = 14°$, and so on. You will find that the cosine wave that can be drawn most closely over these points has an amplitude (or half-height) of about 9°, which is just what is needed for a curvature of 45° per double-helical turn.

When we consider the DNA that is found in our chromosomes, we might expect to find AA/TT sequences where the DNA curvature requires low roll, and GC sequences where the curvature requires high roll. That is precisely what we do find, to a first approximation. Most of the DNA in our chromosomes curves strongly around protein "spools" (Fig. 1.5) for almost two turns of 80 base-pairs each, or for about 160 base-pairs in all, into a flat, left-handed supercoil that resembles part of a telephone cord. The proteins that make up any spool are called "histones," and they will be discussed in Chapter 7. When we examine the DNA sequences that reside in these tight coils that wrap around the protein spools, we find that AA/TT sequences have a higher probability of being in a low-roll position than other DNA, and that GC sequences have a higher probability of being in a high-roll position; but not every AA/TT goes to low roll and not every GC goes to high roll. The preferences of all 160 base-pair steps in this DNA have to balance against one another to attain an optimal positioning of roll angles for *most* of the steps in the sequence; and *some* steps do not fit into the overall pattern.

Some actual data are shown in Fig. 4.10, concerning the preferred locations of different base sequences in curved, chromosomal DNA. To construct these plots, we isolated the DNA from over one hundred different spool–DNA complexes, and then counted how many times N each kind of sequence might be found at any possible location on the histone spool. During the isolation of this DNA, it was trimmed in length from 160 base-pairs down to 145 base-pairs by a special enzyme, in order to mark more precisely where each spool might begin or end. The plots in Fig. 4.10 show data that have been averaged over positions 1–72 and 73–144 of the spool–DNA complex, about the center at 72.5, because these data were found to be the same on both sides of the center.

Fig. 4.10(a) shows the number N of AA/TT steps that were found at each possible location along the path of the long, curved DNA in our many examples. These AA/TT steps may be seen to be located preferentially at positions 5, 15, 25, 36, 46, and 56 from either end of the DNA; such locations are known to be positions of low roll, where the minor groove faces inwards, from an X-ray analysis of the spool–DNA complex. Around the center of the DNA at position 72.5, the periodic pattern is broken, and peaks for AA/TT appear at positions 62 and 72. It turns out that the preference of AA/TT steps for these central positions is not due to DNA curvature, but rather to specific contacts between certain atoms on the protein and the exposed edges of the A/T base-pairs. So we need not worry about this detail; the bulk of the DNA shows roll-angle preferences in accordance with our general scheme.

Fig. 4.10(b) shows, similarly, the number N of GC steps at each possible location along the long, curved DNA. These GC steps may be seen to be located preferentially at positions 11, 21, 31, 41, 52, or about 5 steps away from the preferred locations of AA/TT steps. In other words, they prefer positions of high roll. The mean periodicity, or spacing of peaks, in both the AA/TT and GC patterns is close to 10.2 base-pairs, or one turn of double helix. The preferred helical periodicity of this DNA in other circumstances, when freed from the protein, is a slightly larger 10.6 base-pairs per turn.

Fig. 4.10(c) shows an especially strong pattern of preferences for the trimer AAA/TTT: here two adjoining AA/TT sequences reinforce each other's individual preferences. Thus the AAA/TTT trimer has a strong preference to be located in a position of low roll rather than a position of high roll: the value of N differs by a factor of about 2 between these locations. This is quite a large factor, when you remember that the piece of DNA which is being curved around the histone spool consists of 144 steps or 145 base-pairs. Suppose this piece of DNA contains just one AAA/TTT trimer: then in 2 out of 3 cases, the AAA/TTT will

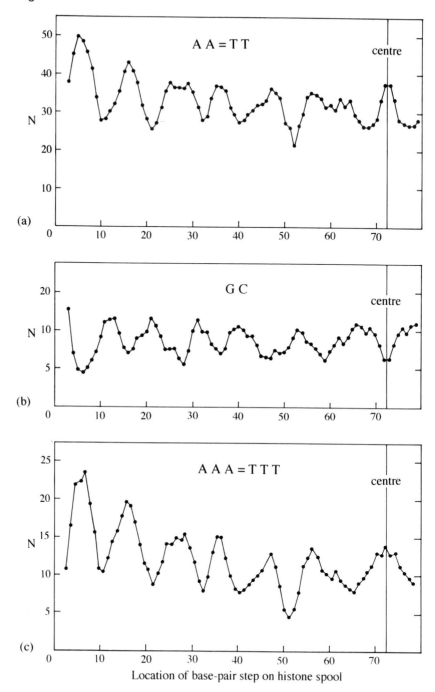

Figure 4.10 Number *N* of various short sequences in curved chromosomal DNA, plotted against their locations on the histone spool (see Fig. 1.5). The overall length of the DNA is 145 base-pairs, but the data could be averaged on both sides of the center at position 72.5, because they were found to be the same on both sides. Over one hundred different DNA molecules were sequenced in order to construct these plots. From S.C. Satchwell *et al.* (1986) *Journal of Molecular Biology* **191**, 659–75.

be located in a position of low roll, due to its own influence in combination with the influences of other sequences. From this example, we can see that roll-angle preferences of particular short (i.e. two- and three-base-pair) sequences in DNA may be rather strong, provided they are not balanced by the opposing preferences of other sequences. Thus, certain AA/TT steps might predispose the DNA to adopt one position on the spool, while certain GC steps might prefer a different position. This is like the story of the husband and wife who always vote for different political parties, Republican and Democrat, or Conservative and Labour; in the end their votes cancel.

When studying pieces of DNA from a chromosome that are longer than the 145 base-pairs examined above, often we see that the spool–DNA complexes align themselves into an ordered array that may extend for 1000–2000 base-pairs. In such cases, the roll-angle preferences within any small region of the DNA, say 100 base-pairs, would have to be very strong to dominate the alignment of the other 900–1900 steps. Alternatively, these 100 steps could "break" the array to yield a less regular structure. In any case, the fit of roll-angle preferences in the DNA on one histone "spool" to those of the DNA wrapped around neighboring spools is a matter of some interest, but it is not yet understood.

Proteins that bind even to a very small region of DNA seem to follow the same rules of DNA curvature. For example, a protein from a bacterial virus, known as the "434 repressor" (because it stops RNA from being made at a site of DNA unwinding in 434 virus), binds to DNA in the following way: it probes deeply into the helix grooves on either end of a 14-base-pair region, and curves the helix moderately in the middle, without making contact with it there, as shown in Fig. 4.11. In this structure, the base-pair steps around bases 6, 7, 8, 9 must adopt a low roll for the DNA to be bent by the protein in a correct fashion, so that the two molecules can fit together properly. Indeed, if you put a low-roll sequence of four bases such as AATT in positions 6, 7, 8, 9 of the DNA, the protein binds 300 times more tightly than if you put a high-roll sequence such as GGCC. In chromosomes, such preferences are averaged out over a long region of DNA so there they appear to be weak; but in small regions of DNA, these preferences are not averaged out, and so they act strongly. In fact, you can calculate from the kind of bases which are found in positions 6–9 of the binding site how tightly this 434 repressor protein will bind to DNA, by using roll-angle preferences taken from the spool–DNA data discussed above. The same physical phenomenon of curvature-preference in DNA is operative in both cases.

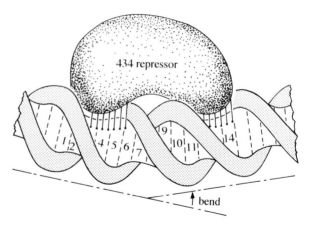

Schematic view of the binding of 434 repressor to DNA. This protein probes for the identities of base-pairs in positions 1–5 and 10–14 of its binding site, and also docks to the overall DNA conformation by testing the ease of curvature in positions 5–9.

But how does the 434 protein bind to DNA in positions 1–5 and 10–14, where it contacts the base-pairs most closely? Is the flexibility of DNA over positions 6–9 sufficient to provide enough specificity for the protein to carry out its function in the cell, of repressing the synthesis of RNA from just one gene on the DNA? For that purpose, the protein needs a specificity of binding to its particular target sequence over other sequences by a factor of 10 000, while the flexibility of DNA provides a factor of 300 at most. The biological mechanism by which this protein blocks RNA synthesis is trivial: it simply binds to the same piece of DNA as that preferred by RNA polymerase, and so it physically prevents the polymerase from starting RNA chains. But the way in which the 434 repressor picks out a single DNA sequence (or just a few sequences) from many others in the bacterial chromosome is not easy to understand.

During the last few years, single crystals of several protein–DNA complexes have been analyzed by X-ray methods, and this has made it possible to study in fine detail the close bonding of different parts of protein molecules to the bases of DNA. The general picture which has emerged from these studies is that the amino acids of the protein can make specific hydrogen bonds with exposed atoms on the sides of base-pairs, or along the "floor" of the major or minor groove in the DNA. A typical interaction of this sort is shown in Fig. 4.12, where a guanine base is shown making hydrogen bonds to an arginine amino acid. There are two contacts of hydrogen atoms on the arginine with oxygen or nitrogen atoms on the major-groove edge of the

guanine ring. If the spatial patterns of electric charge on the
two surfaces fit each other well, so that two or more hydrogen
bonds can form, then there will be a highly specific local
bonding arrangement between the amino acid and the base-pair.
However, it is not completely true to say that the detailed patterns
of amino acid-to-base hydrogen bonding, when summed over
many different amino acids and bases, actually determine where
a protein binds on DNA. For we have seen already that the
flexibility of DNA plays a large part in its binding to the 434
protein, and to the histone–spool proteins, and to other proteins
not discussed here; and this flexibility depends on the DNA
sequence as well.

The picture we need in order to explain all aspects of
protein–DNA recognition is one which involves consecutively
two processes which we shall call "docking" and "probing." By
docking we mean the fitting together, on a large scale, of the
protein and the DNA. If the fit at that scale is good, then the
quality of hydrogen bonding between the contacting zones can
be tested on a small scale; only if this detailed probing is
successful will the overall binding between the protein and the
DNA be highly specific for some particular base sequence. In
our example of the 434 protein, the docking phase demands a
capability of the DNA to bend in a certain way; if the DNA is
too rigid, or prefers to bend in the wrong way, the docking
cannot easily occur. But if the DNA has the right degree of
flexibility, then the base-pairs in the contacting regions 1–5 and
10–14 can probe for the formation of enough hydrogen bonds
with amino acids to let the second stage of binding take place.

Figure 4.12 Hydrogen bonding between a guanine base in DNA and the
arginine amino acid in a protein, as seen in many protein–DNA complexes
(such as that shown in Fig. 4.13).

Another example of the same two-stage process of recognition between DNA and protein is the cutting of DNA by the enzyme DNAase I. This enzyme is very useful for the study of DNA structure, as we shall see in Chapter 8. It can only dock with DNA if the width and depth of the minor groove lie within a certain range, and only after it has docked can it begin to probe the configuration of the sugar–phosphate chain, and make a cut if it finds a specific geometry there. Thus the enzyme must recognize not only a *global* feature (the groove dimensions) but also a *local* feature (the phosphate configuration) if it is to cut the DNA. These two features together provide typically a factor of 1000 in the specificity of cutting some bonds over others.

A third interesting example of protein–DNA recognition was provided recently by the X-ray analysis of a complex between a protein called Zif268 and the sequence to which it binds on DNA. Zif268 is known as an "activator of transcription," or a "transcription factor," because it somehow stimulates the synthesis of RNA from certain genes close to where it binds on the DNA. No one today is certain how this might be accomplished.

Nevertheless, the structure of the complex between Zif268 and DNA is of interest in its own right. The protein contains a series of small, modular units known as "fingers," as shown in Fig. 4.13. Each unit contains a zinc atom which helps to fold the protein chain into a separate domain, and these domains are linked by short segments of the protein chain. Some proteins have as many as 10–20 fingers, but Zif268 has just three. When the Zif268 protein binds to DNA, it tries to attach itself to the DNA base-pairs through a series of arginine-to-guanine hydrogen bonds, as shown schematically in Fig. 4.13. Not shown in this diagram are many other hydrogen bonds which connect phosphate groups from one of the DNA sugar–phosphate chains to various other amino acids on the protein. In fact the three linked fingers form a spiral, which drapes itself along the sugar–phosphate chain on one side of the major groove, and thereby allows arginine amino-acids to make firm contact with guanine bases in 6 out of 9 successive base-pairs. In other words, many contacts of the protein with the phosphates anchor it in place, so that it can contact the guanines.

How does such a zinc-finger protein "dock" to the overall structure of DNA? In this example, the base-pairs of the DNA have moderately high slide, and it may be that the high slide imparts a particular spatial relationship to the phosphates and bases, which then constitutes a three-dimensional docking

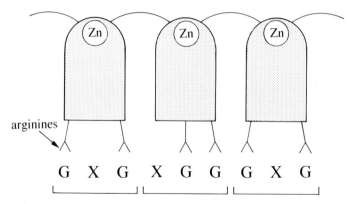

Figure 4.13 Schematic view of three "zinc-fingers" recognizing a particular DNA sequence, as in the X-ray structure of the Zif268 protein with DNA. Not shown are many contacts of the protein with DNA phosphates, or the moderately high slide of the base-pairs.

feature for any individual finger. Alternatively, the high slide may act by increasing the distance of base-pairs from an overall helix axis, as shown in Figs 3.14(b) and 3.15, in which case the three fingers of the protein would have to follow a helical path of greater diameter than before. More data of a precise kind are needed before we can understand fully how this protein recognizes particular sequences in DNA.

In summary, this chapter has been about the twisting and curving of DNA, and how DNA is recognized by proteins according to its twisting and curving and other features. In order for the bases of DNA to be exposed for copying into more DNA or into RNA, the double helix has to untwist. The places where DNA unwinds in Nature, to start the copying process, are often the weakest parts of the double helix. In order for DNA to wrap itself tightly into chromosomes, the double helix has to curve. The curvature spreads itself out over as many base-pairs as possible, in order to keep the roll angle at individual base-pair steps small, so that the bases will stack well onto each other. The sequence of the DNA influences how easily DNA can adopt any given curved shape, by the preferences of different sequences for different roll angles. These preferences apply not only to protein–DNA complexes as part of a chromosome, but also to protein–DNA complexes generally, such as those formed between the DNA and "repressor" proteins. In general, a protein recognizes certain sequences in DNA by first recognizing some large-scale feature of the molecule such as the sugar–phosphate chains (in other words, by "docking"), and then by probing the details of the bases.

We have now reached the halfway point in our book, and we hope that everyone is still with us. We have talked about curving DNA in a plane, but not in three-dimensional space; we have treated twisting and curving as separate subjects, whereas often they are related by the shape and thermal vibration of the DNA. We have not yet discussed how a naked DNA molecule is assembled into a mixture of protein and DNA to make a chromosome, nor how genes are activated within chromosomes by removal of some of the proteins. Lastly, we have not explained any of the experimental techniques which scientists use to probe DNA structure and function, such as X-ray diffraction, nuclear magnetic resonance, and gel electrophoresis. Chapters 5–8 will deal with these and other topics.

Notes

1. See Appendix 1.
2. New work by Susumu Hirose suggests that the protein TFIID unwinds DNA in the vicinity of the TATA sequence by slightly more than one double-helical turn.

Further Reading

Anderson, J.E., Ptashne, M., and Harrison, S.C. (1987) Structure of the repressor–operator complex of bacteriophage 434. *Nature* **326**, 846–52. X-ray crystal structure of a complex between the 434 protein and DNA, showing both "docking" and "probing" modes of binding.

Calladine, C.R. and Drew, H.R. (1986) Principles of sequence-dependent flexure of DNA. *Journal of Molecular Biology* **192**, 907–18. How roll angles in the DNA can change to let DNA bend tightly about a protein.

Drew, H.R., Weeks J.R., and Travers, A.A. (1985) Negative supercoiling induces spontaneous unwinding of a bacterial promoter. *EMBO Journal* **4**, 1025–32. The −10 region of a bacterial promoter, having a sequence of the kind TATA, unwinds more easily than other regions of the DNA at room temperature under torsional stress.

Fairall, L., Rhodes, D., and Klug, A. (1986) Mapping of the sites of protection on a 5S RNA gene by the *Xenopus* transcription factor IIIA. *Journal of Molecular Biology* **192**, 577–91. A good study of how a well-known zinc-finger transcription factor, TFIIIA, binds to its target sequence on the gene that it helps to control.

Gotoh, O. and Tagashira, Y. (1981) Stabilities of nearest-neighbor doublets in double-helical DNA determined by fitting calculated melting profiles to observed profiles. *Biopolymers* **20**, 1033–42. First detailed experimental data showing low heat-stabilities for pyrimidine–purine steps, especially those of the kind TA.

Koudelka, G.B., Harrison, S.C., and Ptashne, M. (1987) Effect of non-contacted bases on the affinity of 434 operator for 434 repressor and Cro. *Nature* **326**, 886–8. Solution data on how the 434 and Cro repressors dock to DNA.

Liu, L.F. (1989) DNA topoisomerase poisons as antitumour drugs. *Annual Reviews of Biochemistry* **58**, 351–75. A summary of drugs used in the treatment of cancer that poison topoisomerase proteins.

Luisi, B.F., Xu, W.X., Otwinowski, Z., Freedman, L.P., Yamamoto, K.R., and Sigler, P.B. (1991) Crystallographic analysis of the interaction of the glucocorticoid receptor with DNA. *Nature* **352**, 497–505. The first crystal structure analysis of a hormone-receptor protein bound to DNA.

Miller, J., McLachlan, A.D., and Klug, A. (1985) Repetitive zinc-binding domains in the protein transcription factor IIIA from *Xenopus* oocytes. *EMBO Journal* **4**, 1609–14. The earliest correct proposal of a repeating structure for the zinc-finger proteins.

Pavletich, N.P. and Pabo, C.O. (1991) Zinc finger-DNA recognition: crystal structure of a *Zif268*–DNA complex at 2.1 Å. *Science* **252**, 809–17. Three-dimensional structure of a transcription factor somewhat like TFIIIA bound to DNA; the structure shows the probing of guanine bases by arginine amino acids, and also high slide in the DNA (although this is hard to see in the pictures).

Satchwell, S.C., Drew, H.R., and Travers, A.A. (1986) Sequence periodicities in chicken nucleosome core DNA. *Journal of Molecular Biology* **191**, 659–75. Different base sequences occupy different locations within DNA as it curves, owing to their preferences for different kinds of roll angle.

Spassky, A., Rimsky, S., Buc, H., and Busby, S. (1988) Correlation between the conformation of *Escherichia coli* −10 hexamer sequences and promoter strength. *EMBO Journal* **7**, 1871–9. More examples to show that easily unwound TATA sequences help RNA polymerase to function.

Bibliography

Giaever, G.N. and Wang, J.C. (1988) Supercoiling of intracellular DNA can occur in eukaryotic cells. *Cell* **55**, 849–56. Confirmation in yeast cells of a theoretical model for the influence of transcription on DNA supercoiling.

Liu, L.F. and Wang, J.C. (1987) Supercoiling of the DNA template during transcription. *Proceedings of the National Academy of Sciences, USA* **84**, 7024–27. A classic paper describing what might happen if RNA polymerase could not rotate freely about the DNA during transcription.

McClellan, J.A., Palacek, E., and Lilley, D.M.J. (1986) (A–T)$_n$ tracts embedded in random-sequence DNA: formation of a structure which is chemically reactive and torsionally deformable. *Nucleic Acids Research* **14**, 9291–309. The unwinding of TATA-like sequences to a twist of 12 base-pairs per turn before a cruciform appears.

Rosenburg, B., VanCamp, L., and Krigas, T. (1965) Inhibition of cell division in *Escherichia coli* by electrolysis products from a platinum electrode. *Nature* **205**, 698–9. Effects of an electric field on bacterial cell growth, and a deduction that a platinum compound might be involved.

Rosenberg, B., VanCamp, L., Trosko, J.E., and Mansour, V.H. (1969) Platinum compounds: a new class of potent antitumour agents. *Nature* **222**, 385–6. Testing of platinum compounds against cancer in mice, with positive results.

Exercises

4.1 Make a physical model to show some of the problems that occur when a polymerase molecule runs along double-helical DNA, as in Fig. 4.3.

To do this, take two equal lengths of rubber tubing, and tie them together firmly at one end. Then, working from that end, arrange the two tubes in the form of a fairly loose right-handed double helix; and when you get to the other end, tie the tubes together there also.

Now insert a stick between the two tubes, near the mid-point of the double helix. Holding an end of the double helix in one hand and the stick in the other, push the stick along the double helix. Observe how the helices tighten up ahead of the stick, and relax behind it—unless, that is, the stick is allowed to rotate as it moves forward along the model.

(This demonstration was suggested by Maxim Frank-Kamenetskii.)

4.2 The base sequence of a single strand of double-helical DNA is given below:

(5′) ACTTAAGGCCCTATATACCTAGACTCGGCGGTAAATTT (3′)

a Underneath it, write out the base sequence of the complementary strand.

b Identify AT-rich and GC-rich regions of the molecule.

c Identify pyrimidine–purine steps.

d Hence identify strong, medium, weak and very weak regions of cross-chain bonding, in the manner of Fig. 4.4.

4.3 The base sequence of a single strand of double-helical DNA is given below:

(5') GCGCCTAGAAATAATACTA GTATTATTTCTAGCCGG (3')

a Underneath it, write out the base sequence of the complementary strand.

b Find a region which can make a "cruciform" formation, as in Fig. 4.5; and draw a picture to show the new pairing.

4.4 A semicircle of garden path around a tree (cf. Fig. 4.7) is made from 10 rectangular slabs, with an individual angle of 20° at each of the 9 "steps" between consecutive slabs.

a How many slabs, arranged in the same pattern, are needed to make a complete circle? Given that the center-to-center spacing of the slabs is 0.6 m, what is the circumference of the circle? And what is the radius of the circle?

b Now do a different calculation on the same curved path, in order to find its radius. First, convert the curvature of 20° per slab into units of radian per meter by the use of the conversions: 1 radian = $180°/\pi$ = 57.3°; 1 slab = 0.6 m. Then calculate the radius of the circle from the formula (which we have not mentioned before, but which is easy to prove): radius of curvature = 1/(curvature).

4.5 Imagine that the four DNA molecules shown in Fig. 4.8 have been made with a curvature of 30° per helical turn, instead of 45° per turn.

Sketch out the corresponding version of each of the four plots (a) to (d) in Fig. 4.9, and list the maximum and minimum values of roll angle R in each case.

4.6 A particular DNA molecule has $T = 36°$ at every step.

a The pattern of roll angles along the length of the molecule is somewhat similar to that shown in Fig. 4.8(c) and Fig. 4.9(c), except that the batches of non-zero roll are only three steps long, and the individual roll angles are smaller. Specifically

$$R_n = (10°, 10°, 0°, 0°, 0°, 0°, 0°, 0°, 0°, 10°)$$

where $n = 0$–9, and the pattern repeats many times.

Compute the curvature of the molecule (in units of degrees per helical turn) by evaluating the sum of 10 consecutive terms

$$R_n \cos (36°n), \quad n = 0 \text{ to } 9.$$

To do this, make a table of 10 rows and 4 columns, with the following entries:

Column 1, values of n;
Column 2, values of $\cos (36°n)$;
Column 3, values of R_n;
Column 4, values of $R_n \cos (36°n)$.

Obtain the required answer by taking the sum of all entries in Column 4.

b Repeat the calculation for

$$R_n = (0°, 0°, 0°, 0°, 10°, 10°, 10°, 0°, 0°, 0°)$$

c Repeat for

$$R_n = (0°, 0°, 0°, 0°, -10°, -10°, -10°, 0°, 0°, 0°)$$

d Repeat for the case (which is somewhat similar to that of Figs 4.8(d) and 4.9(d), and to Exercise 4.5):

$$R_n = 6° \cos (36°n), \quad n = 0 \text{ to } 9$$

e Repeat for

$$R_n = 6° \cos (36°n) + 4°$$

4.7 Fig. 4.12 shows the hydrogen bonding between a guanine base of DNA and an amino acid arginine on the peptide chain of a nearby protein. The diagram below shows an amino acid asparagine in a similar way. Would asparagine form hydrogen bonds better to guanine or to adenine, in the major groove of DNA, as in the style of Fig. 4.12? See Fig. 2.11 for chemical formulas.

asparagine

CHAPTER 5

Curving in Three Dimensions

So far in this book we have said a lot about the structure of DNA as a double helix, and the way in which different kinds of double helix can result from the different geometries with which base-pairs stack on one another. But in most chapters our DNA, seen as a long rod or thread, has been straight. Only in Chapter 4 did we consider a molecule of DNA which follows a curved path; and the curve there was in a plane, like a garden path. We have now reached the stage where we must explain the curvature of DNA in three dimensions.

A DNA helix often proceeds through space as a three-dimensional spiral, rather than as a straight line or as a plane curve. The spiral path of DNA is usually described as a "superhelix" or "supercoil." This is because we know that the DNA thread itself has a local helical, twisted structure even when it is straight. The qualitative aspects of DNA curvature in three-dimensional space are not difficult to understand, although we shall need to be careful to distinguish between the twisting of the path of the DNA as a whole, and the twisting of sugar–phosphate chains on a local scale. But you may find that some of the quantitative and mathematical aspects of the subject are hard to grasp. Do not worry if you find the mathematics rather heavy going: be content to appreciate the qualitative aspects of three-dimensional curvature.

In Chapter 4, we explained how DNA can curve in a plane to make a circle. There, curvature k was defined as the angle turned for a given length of DNA. For example, k might be $1°$ per base-pair or, equivalently, $10°$ per double-helical turn of 10 base-pairs. When DNA curves in three dimensions, our simple definition of curvature k still holds for any small segment of its path, such as one helical turn or 10 base-pairs; but in practice the DNA often departs from the local plane of curvature after two or more turns. In such cases, the plane of curvature twists by some angle *t* as the DNA advances through space, and so the DNA coils into three dimensions.

There is an easy way to understand this point. Consider a little man in an airplane (Fig. 5.1). In Fig. 5.1(a) he flies loop-the-loops, which are circular paths in a vertical plane, like the path of a car in a carnival ferris wheel; and he does so by pointing the nose of his plane either up or down, using his joy-stick to tilt the elevators on the tail of the plane. This is pure curvature *k*, since his path through the air curves by some angle *k* per unit length of loop. In Fig. 5.1(b), on the other hand, he flies straight ahead but in a narrow spiral, by pushing the joystick either to the left or to the right, and so activating the ailerons on the wings. The path of the aeroplane is now one of pure twist *t*, since the plane rotates about its long axis by some angle *t* per unit length of advance through the air. The twist *t* can be either plus or minus, depending on whether he turns the wheel to the right or to the left, respectively. After a while, the man gets dizzy from flying in a straight, twisted path, and he decides to combine the two motions of *k* and *t*. So he turns the nose of his 'plane both up and to the left at the same time; and he now flies in a broad spiral through the air, like the path of the wire in a coiled bed-spring. It is the *combination* of *k* and *t* that provides for a broad spiral path; *k* alone gives a plane circle, while *t* alone gives a narrow spiral like a twisted ribbon.

This story gives us a good initial understanding of how curvature *k* and twist *t* influence the path of DNA through space. Let us now ask, first, how *k* and *t* influence the shape of any small segment of DNA; and second, how these small segments can be joined together to make different paths through space, like those flown by our friend in the airplane.

We can represent any small segment of DNA as in Fig. 5.2(a) by a short cylinder with the sugar–phosphate chains along its outside. This particular segment contains exactly one double-helical turn of 360°, or 10 base-pair steps. The first and last base-pairs, numbers 0 and 10, can be drawn as black rectangles on the two ends of the cylinder, to demonstrate that they are parallel to one another, and to the vertical in the drawing. Only one of these two black rectangles can be seen in the perspective of Fig. 5.2(a), so you have to imagine that there is another one on the far end of the cylinder.

An even simpler version of the same thing is shown in Fig. 5.2(b), where the turn of DNA is now represented by a solid block. The two ends of the block are perfectly aligned,

(a)

(b)

Figure 5.1 Our friend in the airplane explains to us about curvature k and twist t. Curvature k alone as in (a) makes the 'plane fly in vertical loop-the-loops. Twist t alone as in (b) makes the 'plane fly in a narrow spiral. Only the combination of curvature k and twist t produces a broad spiral.

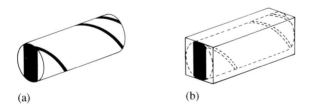

(a)   (b)

Figure 5.2 (a) One complete turn of double-helical DNA, showing two sugar–phosphate chains along the outside, and a black rectangle as the first base-pair: the whole structure has been drawn as a cylinder. (b) The same as in (a), but now the DNA has been encased in a semitransparent solid block.

because the DNA has twisted by 360° along its axis in going from one end to the other; and the block is straight as in (a). You can imagine that the cylinder of (a) is converted into the block of (b) by first glueing squares to the ends of the cylinder, and then filling in the space between the squares with some sort of semi-transparent jelly which enables us to see, darkly, the sugar–phosphate chains buried inside. From now on we shall mainly disregard the twisting of these strands within the block, and will usually draw the block as in Fig. 5.3(a), without any hint of what is actually hidden inside it. In this figure both the curvature k and the twist t of the block are precisely zero, and all six faces of the block are plane rectangles.

Figure 5.3(b) shows a block that has curved by 10° to the left, or equivalently by $k = 1°$ per step over 10 steps. It looks something like a banana. The upper surface is obviously curved, while the shaded surface might form part of a cylinder. Presumably, the roll angles R of the DNA within the block have changed on going from (a) to (b), so that the best-fit cosine wave to these roll angles, as described in Chapter 4, now sweeps out an angle of 10°. The twist t is still zero, because the two ends of the block remain aligned: note that the closely spaced lines on the shaded surface are all strictly parallel to each other.

Figure 5.3(c) shows a block that has twisted in a negative or left-handed sense by 10° from end to end, or equivalently by $t = -1°$ per step over 10 steps. Presumably, the local twist angles T between successive base-pairs in the DNA have decreased from $T = 36°$ to $T = 35°$ on average, in going from (a) to (c). Thus, the 10 steps of hidden DNA now twist only by a total of 350° rather than 360°, leaving a deficit of $350° - 360° = -10°$; this is what shows in the picture. In (c) there is no curvature: the top surface, though now twisted, is still rectangular.

Lastly, Fig. 5.3(d) shows a block that is both twisted and

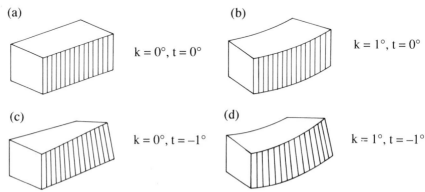

(a) $k = 0°, t = 0°$

(b) $k = 1°, t = 0°$

(c) $k = 0°, t = -1°$

(d) $k = 1°, t = -1°$

Figure 5.3 Three possible ways by which the solid block shown in Fig. 5.2(b) can change its shape, by the application of uniform curvature k or twist t. In this figure, curvature k and twist t are given in units of degrees per base-pair. Since there are 10 base-pair steps, the total curvature is 10° in (b) and (d), while the total twist is 10° in (c) and (d).

curved; it looks like a slightly-twisted banana. Both roll R and twist T have changed at the base-pair level to produce this result, in the same ways as described for parts (b) and (c) previously; but now R and T have both changed simultaneously.

We have used the word "twist" in two senses here, with two symbols t and T. Lower-case twist t is the difference in total twist from 360° after one unit-length of helix. It amounts to $-10°$ after 10 steps in (c), as explained above. Upper-case twist T is the twist of any base-pair step locally. For example, $T = 36°$ in both (a) and (b), but $T = 35°$ in (c) and (d). It follows, therefore, for any unit of n base-pair steps with uniform twist T:

$$t = nT - 360°$$

Here $t = (10 \times 35°) - 360° = -10°$ in both (c) and (d) of Fig. 5.3. What do we mean by a "unit of n base-pair steps"? Basically, we mean that the internal structure of the DNA is identical in each successive set of n steps—as it is likely to be, for example, if the *sequence* of the DNA repeats every n steps.

In summary, curvature k and twist t can describe the shape of any small segment of DNA on a local scale. These values of k and t come from variations in base-step roll R and twist T, respectively. We explain below how to calculate k and t from given values of roll R and twist T, for the base-pair steps of practically any segment of DNA.

Now you must use your imagination. What happens if we take many identical blocks of the kind shown in any one part of Fig. 5.3, and join them together, end-to-end, over a long distance? Using blocks all of type (a), we would get a straight, untwisted rod; using blocks of type (b), we would get a plane circle; while the blocks of type (c) would build a straight but twisted rod. Finally, the blocks of type (d) when joined together would make a broad spiral that we might call a "superhelix," if we remember the DNA hidden within it. Such a spiral will rotate counter-clockwise as it goes forward, and be left-handed, because the sign of twist t is negative. If t were positive, we would get a right-handed superhelix. This would happen, for example, if the local twist T became 37°, so that $t = nT - 360° = (10 \times 37°) - 360° = +10°$.

In fact, we do not have to rely entirely on our imagination for these constructions. We can *calculate* the path of the DNA through space for different values of k and t, by using certain geometrical formulae. Therefore, let us now derive these formulae.

A typical left-handed superhelix is shown in Fig. 5.4(a). This superhelix might be made by joining together many blocks like the one shown in Fig. 5.3(d). It winds uniformly like a ribbon about a vertical cylinder of radius r, at an angle α with respect to the horizontal at any point. It makes a left-handed spiral as it winds round the cylinder. It turns by 360° around the vertical axis of the cylinder after a vertical distance p, which is known as the superhelical "pitch." Finally, it has a contour length, not indicated in the picture, of N^* base-pair steps in one 360° turn. If an ant were to crawl along the path of the superhelix for any full 360° turn around the cylinder, and then measure over how many steps he had crawled, that would be N^*.

The usual way to analyze these parameters r, α, p, and N^* is to imagine that we can "unroll" the cylinder onto a piece of paper, in the same way that you can unroll the cardboard cylinder on which paper towels are wrapped. This is called a

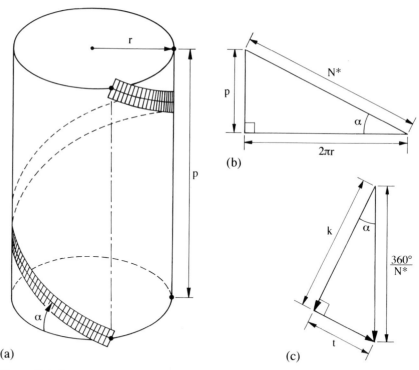

Figure 5.4 Curvature and twist in a spiral ribbon. In (a), a left-handed spiral ribbon of pitch angle α and radius r goes through one turn every p units of distance along the cylinder. In (b), the surface of the cylinder has been unwrapped on a smaller scale, to show geometrical relations among α, r, p and contour length N^*. In (c), the vector sum of curvature k and twist t, on a very-small scale, gives the amount by which each step of the ribbon rotates around the axis of the cylinder. Note that triangle (c) is similar to triangle (b).

"cylindrical projection," and it is shown, to a smaller scale, in Fig. 5.4(b). The path of the superhelix becomes the diagonal of a right-angled triangle, while the pitch p and the circumference $2\pi r$ make up its other two sides; they are related by the angle α. It is a simple matter to find all of N^*, p, r, and α from any two of these four parameters; but we shall not do so here.

All of this is very well-known geometry; but the next part is not so well-known. Consider any small part of the superhelical path, say just 2° of the total 360° for each turn around the cylinder. For example, when $N^* = 180$ base-pairs, any single base-pair makes up $360°/N^* = 2°$ of rotation about the circumference. In this case, the values of curvature k and twist t for the spiral ribbon are related to $360°/N^*$ as shown in Fig. 5.4(c). Curvature k is the cosine component of the advance by 2°, while twist t is the sine component. This result follows from a branch of mathematics known as "differential geometry," for the special case of a ribbon wrapped about a cylinder. In simple terms, the curvature k tells how far the superhelix should curve around the cylinder in the manner of Fig. 5.3(b), and t tells us how much the ribbon should twist up the cylinder in the manner of Fig. 5.3(c). The sum of the two effects (it is actually a vector sum) gives the total rotation about the axis of the cylinder.

Several interesting equations follow from the triangle of Fig. 5.4(c):

$$\tan \alpha = t/k$$

$$k = (360°/N^*) \cos \alpha$$

$$t = (360°/N^*) \sin \alpha$$

$$(360°/N^*)^2 = k^2 + t^2$$

We can now relate the diagrams in Fig. 5.4(b) and (c) to one another, and thereby get the overall shape of the superhelix from any given values of k and t. By similar triangles we have

$$p/N^* = t/(360°/N^*)$$

Substituting for N^* from the last of the four equations above, we obtain

$$p = 360° \, t/(k^2 + t^2).$$

Similarly,

$$2\pi r = 360°k/(k^2 + t^2)$$

Let us consider as a specific example a case where $k = 1°$ and

$t = -1°$ per base-pair. Then $\alpha = -45°$, meaning that the superhelix is left-handed as in Fig. 5.4(a), since t is negative. Also, contour length $N^* = 360°/\sqrt{2} = 255$ base-pairs, while pitch $p = 180$ base-pairs. Finally, the circumference of the cylinder $2\pi r = 180$ base-pairs, so radius $r = 29$ base-pairs. The final form of the superhelix is similar to that shown in Fig. 5.4(a), except that the angle α is larger.

In this particular example k and t both had the same magnitude. In a case where t is small compared to k, the formulas give us a low value of α; this means simply that the superhelical coil is almost flat, like a circle going round and round the same path. On the other hand, when k is small compared to t, the formulas give $\alpha \approx 90°$. This means that the superhelix takes the form of a highly extended spiral, almost like a straight line but with a small superhelical "wobble."

All of this seems very straightforward, once we have learned how to use the formulas. But how can we get k and t from the sequence of the DNA? Unless the base sequence is very regular, k and t will vary from one double-helical turn to the next. In that case, the path of the DNA will not describe a regular superhelix, and so the formulas which we have just derived will be useless. But if the sequence of the DNA does repeat exactly, or almost exactly, once every double-helical turn, then the DNA will form a regular superhelix and our formulas will hold good.

Fortunately, scientists have done many experiments on the structure of "repeating-sequence" DNA, because they can make it rather easily. They synthesize chemically one small part of the DNA, say 10 base-pairs of a defined sequence, and then they join these units together to make a long polymer, just as we imagined when we were studying Fig. 5.3. Having made such repeating-sequence DNA, they can study its structure by the methods of electron microscopy and gel electrophoresis. Electron microscopy shows the shape of the DNA directly, although at low resolution, while gel electrophoresis measures indirectly the "apparent volume" of the DNA cylinder. We discuss this method of analysis in Chapter 8, but for the present we may say that the curved DNA shown in Fig. 5.4(a) could be enclosed in a cylinder of larger volume than for straight DNA of the same length; and so in a gel, curved DNA would come into contact with more gel fibers, and hence go more slowly. This is precisely what is observed in experiments with gels.

Still, it is nice to see the DNA directly, without having to worry about indirect measurements of its volume by gel electrophoresis. For that reason, it is sensible to study repeating-

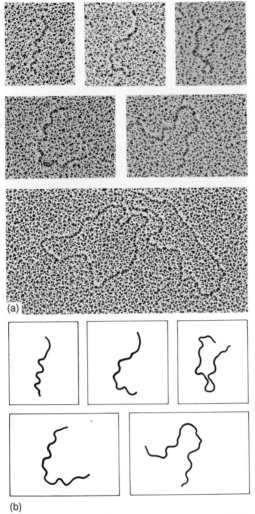

(a)

(b)

Figure 5.5 Electron micrographs of repeating-sequence DNA. The first five frames in (a) show an assortment of pictures of repeating-sequence, curved DNA of size about 1000 base-pairs. The last frame in (a) shows two DNA plasmid or ring molecules of size 3000 base-pairs, that have been included as controls. In (b), approximate tracings of the curved DNA molecules from (a) are presented. Courtesy of Margaret Mott.

sequence polymers by electron microscopy. Some pictures of curved, repeating-sequence DNA as obtained by this method are shown in Fig. 5.5. The first five frames in Fig. 5.5(a) show a collection of similarly curved DNA molecules of size about 1000 base-pairs. They lie in various shapes on the support "grid," having been flattened into two dimensions in preparation for microscopy. The last frame in Fig. 5.5(a) shows two DNA rings or "plasmids" of length 3000 base-pairs, of ordinary (i.e. not repeating-sequence) DNA, as controls for the appearance of DNA with little or no curvature; they were prepared for microscopy under identical conditions. Figure 5.5(b) shows the outlines of curved DNA molecules from (a). Each of the curved DNA molecules has a wiggly, snake-like appearance as compared to the two DNA plasmids. The "wiggle" of the curves on the photographs can be fitted by a sine-wave of contour length N^* = 533 Å and pitch p = 385 Å, as an average over many molecules. The amplitude of the "wiggle" is obviously unreliable as a measure of the original three-dimensional structure, because the DNA superhelix was flattened significantly when placed onto the grid for microscopy; but the values of N^* and p should be representative of the original conditions in solution. From the values for N^* and p, and the triangle shown in Fig. 5.4(b), we can calculate that $\alpha = \pm46°$, while radius $r = 59$ Å before flattening onto the grid. The length per base-pair in these pictures has been measured as 3.0 Å, so we can express the dimensions of the DNA superhelix in units of base-pairs as N^* = 178 base-pairs, p = 128 base-pairs, and r = 20 base-pairs.

Figure 5.5 is a good piece of experimental evidence about the shape of curved, superhelical DNA. But can we calculate the values of N^*, p, and r independently, directly from the base sequence of the DNA, by using a suitable theory? It is easy to determine from the superhelical parameters listed above that k = 1.40° per base-pair and $t = \pm1.45°$ per base-pair; but how can we derive k and t values from the base sequence? Note that the electron microscope pictures do not indicate whether the original superhelix was left-handed or right-handed, and so we do not know from the experiment whether t is positive or negative.

The sequence which was used to make these polymers was of the kind ... AAAAAANNNNAAAAAANNNN ..., where N = C or G, mainly. This sequence repeats once every 10 base-pairs, approximately, as AAAAAANNNN. Now, it is known that the average local twist T for an AA step is close to $T = 35°$, while $T = 34°$ for other steps such as NN, NA, or AN. The slight difference in twist probably comes from an extra hydrogen bond across the major-groove side of the AA step, as shown in Fig. 3.6. Therefore, we can calculate the overall twist t as $(5 \times 35° + 5 \times 34°) - 360° = 345° - 360° = -15°$ per 10 steps, or $t = -1.5°$ per base-pair. Not bad! We have already come close to the value for t of (plus or minus) 1.45° which was obtained by electron microscopy. Both of the values for T at particular steps, which were used in the calculation above, had been determined by gel electrophoresis or other techniques several years before the electron microscope pictures were taken.

But how can we calculate the curvature k from the base sequence? The roll angles R are thought to be close to $0°$ for an AA step, as against $R = +3.3°$ for the others as an average: again, these values come from gel electrophoresis, X-ray crystallography, or other techniques. Such values of base-step roll are plotted in Fig. 5.6 against the step number, for this particular repeating sequence. We must fit a cosine wave to the values of R plotted in Fig. 5.6, in order to determine the curvature k. As explained in Chapter 4, one could draw a "best-fit" cosine wave over the points in Fig. 5.6 by eye, in order to get a satisfactory approximate solution (see the broken line in the diagram), but really we would prefer to have a more systematic way of doing the calculation.

The most accurate method is to take what is known as the "Fourier transform" of the roll angles over any period of steps 0–9 in the base sequence. This is just like what we did in Chapter 4, except that here we show the full mathematics. It doesn't matter which step you assign to be step 0, or which step you assign as step 9, so long as all 10 steps of the repeated sequence (0, 1, 2. . . ., 9) are counted once.

First, you must evaluate two sums, which are the correlations of the roll angles R with either a sine or a cosine wave; for this purpose it is convenient to write R_n for the roll R at any step n:

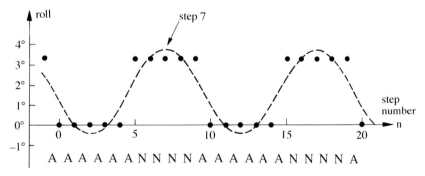

Figure 5.6 A plot of expected roll angles for the repeating-sequence, curved DNA studied by electron microscopy in Fig. 5.5. The curvature k may be calculated as $10.7°$ over any A_6N_4 repeat by taking a Fourier transform of these roll angles, as indicated by the broken line. Roll angles are most positive in the Fourier transform at step 7, where the minor groove of the DNA lies along the outside of the curve.

$$\text{first sum} = \sum_{n=0,9} R_n \sin(36°n)$$

$$\text{second sum} = \sum_{n=0,9} R_n \cos(36°n)$$

Thus, using the numbering of steps shown in Fig. 5.6, we find that:

$$\text{first sum} = 0° + 0° + 0° + 0° + 0° + 3.3° (0.0) + 3.3°(-0.59)$$

$$+ 3.3° (-0.95) + 3.3° (-0.95) + 3.3° (-0.59)$$

$$= -10.2°$$

$$\text{second sum} = 0° + 0° + 0° + 0° + 0° + 3.3° (-1.0) + 3.3° (-0.81)$$

$$+ 3.3° (-0.31) + 3.3° (0.31) + 3.3° (0.81)$$

$$= -3.3°$$

Then the curvature k is simply given by the square root of

$$k^2 = (\text{first sum})^2 + (\text{second sum})^2 = (-10.2)^2 + (-3.3)^2 = 115;$$
$$\text{so } k = 10.7°$$

Our calculated value of $k = 10.7°$ for 10 steps, or $1.07°$ per base-pair, compares fairly well with the value of $k = 1.4°$ per base-pair as determined from the electron microscope pictures of the overall shape. We could also calculate from this theory that our repeating-sequence DNA should make a left-handed superhelix of contour length $N^* = 195$ base-pairs and radius $r = 61$ Å, using the theoretical values of k and t. These are again close to the experimental values of $N^* = 178$ base-pairs and $r = 59$ Å, given above.

Before we go on, let us consider two more features of the Fourier transform: its amplitude and its phase. First, the amplitude of the best-fit cosine wave in most of our examples is equal to the curvature k divided here by 5.0, which is the sum of $\cos^2\theta$ over steps 0–9. Thus, in Fig. 5.6, the amplitude or half-height of the dotted line is simply $10.7/5.0 = 2.1°$. We can say then that the total variation in roll R, for the best-fit wave, is $2 \times 2.1° = 4.2°$. This wave goes from a peak at $R = +3.75°$ to a trough at $-0.45°$ about a mean of $+1.65°$. Second, the phase of a Fourier transform tells us where the best-fit cosine wave is located in a left-to-right sense, relative to the numbering of steps in the sequence. It can be calculated as the

arctangent of the ratio (first sum/second sum)

This gives, in the case above, $\arctan(-10.2/-3.3) = 252°$. This shows us that the origin of the best-fit cosine wave, where its value is most positive, lies 252° to the right of step 0 in the sequence, or in Fig. 5.6 at step (252°/36°) = 7 (see the arrow). In other words, the base-step roll R is the most positive at step 7, and least positive (or most negative) at steps 2 and 12, which are 180° out of phase in either direction. In molecular terms, this means that the minor groove of the DNA lies along the outside of the curve at step 7, and along the inside of the curve at steps 2 and 12.

Suppose we had made another, slightly different DNA molecule with repeating sequence, but this time with a repeat once every 11 base-pairs rather than every 10. For example, AAAAAANNNNN would have an 11 base-pair repeat, as compared to the 10-base-pair repeat for AAAAAANNNN studied above. Then we could calculate the overall twist t by adding up the local twist angles T over 11 steps, to yield $t = (5 \times 35° + 6 \times 34°) - 360° = 379° - 360° = +19°$. Our new superhelix would thus be right-handed, since t is positive. We could calculate the curvature k by taking the Fourier transform of the roll angles over 11 steps, at intervals of 360°/11 = 32.7°, using cosines and sines of angles (32.7° n), to yield $k = 11.6°$. Here, we have calculated both t and k for an 11-step repeat. In units of base-pairs, t and k would be $t = +19°/11 = +1.73°$ and $k = 11.6/11 = 1.06°$. Finally, from these values of t and k we could calculate the dimensions of the right-handed superhelix, by use of the formulas of the preceding section.

Table 5.1 lists values of k and t as calculated for a series of DNA molecules of repeating sequence, which includes the two sequences already mentioned, namely A_6N_4 and A_6N_5. These calculated values of k and t have then been used to calculate the superhelical parameters α, N^* and r for each sequence. Looking first at the two central columns for k and t, we can see clearly that if we had made a 9-base-pair repeat such as A_6N_3, or a 12-base-pair repeat such as A_6N_6, then twist t would have become much larger in magnitude than curvature k. In the case of A_6N_3, you would get $t = (5 \times 35° + 4 \times 34°) - 360° = 311° - 360° = -49°$ per repeat, which gives $t = -49°/9 = -5.44°$ per base-pair. But the curvature k is still only 1.06° per base-pair, and so the pitch angle $\alpha = \arctan (t/k)$ becomes $\arctan (-5.44/1.06) = -79.0°$. This means that A_6N_3 would form a highly elongated, left-handed superhelix of contour length $N^* = 65$ base-pairs and radius $r = 2$ base-pairs. This calculation leads us to the conclusion that we can only expect to find DNA molecules in the form of a *broad* spiral when the DNA sequence-repeat is close to 10 or

11 bases, so that the twist t is low. Of course, if twist $t = 0$ exactly, then the DNA forms a plane circle of contour length $N^* = 360°/k$; and many naturally occurring DNA molecules form plane circles.[1]

How could we calculate k and t if the base sequence was more complicated than A_6N_4 or the related sequences discussed above? Previously we assigned roll R and twist T values only to AA/TT and "other" steps. Thus, we set $T = 35°$ for AA/TT steps but $T = 34°$ for "others." This model is clearly too simple to account for all of the possible arrangements of base-pairs in DNA, but it serves as a first approximation for the case of free DNA in solution, or on an electron-microscope grid. As soon as more data are obtained about the precise shapes of DNA in solution or on a microscope grid, for many different base sequences, it should be possible to assign more accurate values of R and T to every kind of step.

It should be emphasized that free DNA in solution has, in general, little intrinsic curvature. Such curvature probably amounts at most to about 10° or 15° per helix turn; and this level of curvature seems to be reached only for certain base sequences such as those listed in Table 5.1. In close association with positively charged proteins, however, DNA normally curves much more strongly than this, often by as much as 40–50° per helix turn, as described in Chapter 4. The DNA has to deform to fit into the shape required by the protein, since this high level of curvature requires a systematic variation in roll angles of about plus or minus 9°, as we found earlier. In such cases of enforced curvature, a typical dinucleotide step which adopts $R = +3°$ under stress-free conditions in solution might be required

Table 5.1 Calculated superhelical parameters α, N^*, and r (see Fig. 5.4) for DNA of various repeating sequences

Sequence	Repeat	k (°/bp)	t (°/bp)	α (°)	N^* (bp)	r (bp)
A_6N_2	8	1.00	−10.38	−84.5	34.5	0.5
A_6N_3	9	1.06	−5.44	−79.0	65.0	2.0
A_6N_4	10	1.07	−1.50	−54.5	195.4	18.1
A_6N_5	11	1.06	+1.73	+58.5	177.4	14.8
A_6N_6	12	1.03	+4.42	+76.9	79.3	2.9
A_6N_7	13	0.99	+6.69	+81.6	53.2	1.2

The curvature and twist (k, t) were calculated from the base sequence according to angles of roll R and local twist T given in the text.

to change to $R = +9°$ or $-9°$, when the DNA wraps tightly around a protein.

Yet in principle, if we are given a DNA sequence, and we have a table of allowable ranges of roll R for each type of dinucleotide step, we can still test for the ability of DNA to wrap tightly around a protein, by trying to establish a best-fit cosine wave of amplitude 9° through the allowable range. Figure 5.7(a) shows a hypothetical example of this sort, where a DNA molecule is perfectly straight in solution, but can curve around a protein when required to do so. The DNA in this case

Figure 5.7 The ability of DNA to curve tightly about a protein depends on the allowable ranges of roll R at its base-pair steps. In (a), allowable ranges are $R = 0°$ to $-6°$ for steps 0, 1, 2, 3, 4, but $R = 0$ to $+6°$ for steps 5, 6, 7, 8, 9. These roll angles may all remain equal to 0° for DNA free in solution; but when the DNA curves about a protein, many of them must switch to $-6°$ or $+6°$, in order to match the cosine wave of amplitude 9°, shown as a broken line, which coincides with the Fourier term of period 10. This DNA can bend only in one direction, since the best-fit cosine wave has a unique phase. In (b), allowable ranges of R are $-6°$ to $+6°$ for all steps. Now the DNA can bend in any direction, since the same cosine wave can fit the new roll angles with any left-to-right phase.

contains two kinds of dinucleotide step, one with an allowed range of roll $R = 0°$ to $-6°$, and the other with an allowed range of $R = 0°$ to $+6°$. In free DNA, the roll angles are all equal to zero for both kinds of step; and so a horizontal line can be drawn through these points, which indicates that the curvature k must be zero. But when such DNA is forced to bend around a protein, its sequence of roll values must approximate a cosine wave of amplitude 9°. In Fig. 5.7(a), a best-fit cosine wave of amplitude 9° has been drawn as a broken curve; it is clear that most of the steps must adopt their extreme roll values of $\pm6°$ in order to fit the curve. The DNA steps whose minor-groove edges lie along the inside of the curve are those in the middle of each batch (at steps 2 and 12) where $R = -6°$. Any other phasing of this cosine wave would not fit the allowable roll values so well.

In summary, this particular kind of DNA, which is typical of many DNA molecules found in biology, can either be straight, or else it can curve in one specific direction when wrapping around a protein. In this respect, a typical DNA molecule is just like your finger or your knee: it can only bend in one direction and in one plane. On the other hand, if we want to construct a piece of DNA that can bend with equal ease in all directions, we need to include an allowable range of R values from $-6°$ to $+6°$ at every step (or, for example, from $-2°$ to $+10°$). Then the required cosine wave can be fitted at any phase, as shown in Fig. 5.7(b), and the DNA can bend in any direction quite easily.

How can we find out which kinds of dinucleotide step have high or low allowable ranges of roll R, under the bending stress applied by a protein? Some experimental data on this subject have already been presented in Fig. 4.10. Steps such as AA or TT, which prefer to occupy the low-roll locations of DNA when it is wrapped around a histone spool, presumably find it difficult to adopt a high-roll configuration. The preference for low roll is seen to be even stronger for sequences such as AAA; this fits in with the idea that such sequences are relatively rigid, with $R = 0°$. Indeed, it is well-known that DNA molecules containing more than about 50 AA or TT steps in a row cannot easily wrap around a histone spool, because they cannot curve to the degree required. Similar considerations apply to the step GC, which prefers to occupy the high-roll locations of DNA wrapped around a histone spool; it might be assigned an allowable range of R from $+5°$ to $+10°$. Finally, it is thought that steps TA and AT, among others, are relatively flexible to changes in roll, and that they can adopt the full range of R from $-9°$ to $+9°$. Thus, sequences such as TATATATA can wrap with any phase (or angular setting) around a histone spool.

Although we have used the same geometrical ideas in relation both to the intrinsic curvature of repeating-sequence DNA in solution, and also to the enforced curvature of DNA around a protein spool, it cannot be stated too strongly that the curvature of DNA about a protein is very different in magnitude from its curvature when it is free in solution. Most DNA sequences are not curved intrinsically, when they are free in solution, but they often adopt preferred, highly-curved shapes when they bend around a protein, owing to the different kinds of roll-angle flexibility at different sequences.

In Fig. 5.7 we were discussing the simple two-dimensional curvature of DNA in a plane, which is well-understood. But how does the three-dimensional curvature of DNA into superhelical shapes relate to its role in biology? In principle, there are two ways by which this could happen. First, the three-dimensional shape of the DNA might affect how well it binds to different proteins in the cell. For example, one protein may prefer that the DNA should wrap around it as a left-handed superhelix, while another might prefer a right-handed superhelix. Again, one protein may prefer to bind the DNA in the form of an extended superhelix, while another might prefer a flat superhelix. Furthermore, the three-dimensional shape of DNA might affect how it vibrates in solution, in response to the thermal motion of water molecules. It is well-known that things vibrate differently according to their shapes. For example, small molecules such as carbon dioxide (CO_2) and nitrogen dioxide (NO_2) vibrate differently because one is linear whereas the other is bent; and so they show different infrared spectra. Also, the nature of these vibrations or fluctuations may be influenced by any local patches of high flexibility within the DNA, which are present on account of particular base sequences, as shown in Fig. 5.7(b).

Unfortunately, although the theoretical principles are straightforward, chemists and biologists have not yet collected any clear experimental data on how the three-dimensional shape of DNA, over a large scale, affects its protein binding or its thermal fluctuation. The two-dimensional curvature of DNA into a plane has been implicated in many processes, but not yet the three-dimensional curvature into superhelices. All that we know today about the three-dimensional curvature of DNA is what we have learned from physical studies such as electron microscopy and gel electrophoresis. Here is a field where there is plenty of room for improved understanding.

Note

1. Yoshifumi Nishimura has independently derived very similar three-dimensional spiral structures for repeating-sequence DNA, in unpublished work.

Further Reading

Benfante, R., Landsberger, N., Tubiello, G., and Badaracco, G. (1989) Sequence-directed curvature of repetitive *Alu* I DNA in constitutive heterochromatin of *Artemia franciscana*. *Nucleic Acids Research* **17**, 8273–82. Electron-microscope pictures of long, plane-circular curved DNA from a brine shrimp.

Calladine, C.R., Drew, H.R., and McCall, M.J. (1988) The intrinsic curvature of DNA in solution. *Journal of Molecular Biology* **201**, 127–37. All relevant equations for the shape of curved DNA in three dimensions, in terms of its curvature and twist.

Calladine, C.R., Collis, C.M., Drew, H.R., and Mott, M.R. (1991) A study of electrophoretic mobility of DNA in agarose and polyacrylamide gels. *Journal of Molecular Biology* **221**, 981–1005. Pictures of long, superhelically curved DNA such as those shown in Fig. 5.5, by electron microscopy.

Griffith, J., Bleyman, M., Rauch, C.A., Kitchin P.A., and Englund, P.T. (1986) Visualisation of the bent helix in kinetoplast DNA by electron microscopy. *Cell* **46**, 717–24. An early electron microscope study of short, curved DNA.

Exercises

5.1 Any ribbon with uniform curvature k and twist t will generate a spiral or helix. The characteristic shape of the helix may be calculated from k and t by use of the formulas given on p. 99 for pitch angle α, radius r, pitch p, and contour length N^* of one complete turn (see also Fig. 5.4). By convention, k, r and N^* are always positive; while t, p, and α are all positive for a right-handed helix, but negative for a left-handed helix.

Compute values of α (in degrees) and r, p, N^* (in base-pairs) for the four sets of k and t specified below. Which helix has the smallest diameter, and which has the largest? Which helices are left-handed, and which are right-handed?

helix	k (°/bp)	t (°/bp)
a	2	−2
b	2	2
c	2	0.2
d	0.2	2

5.2 The helical geometry of a uniformly curved and twisted ribbon (Fig. 5.4) may be used to provide an approximate calculation for the situation shown in Fig. 3.14(c), where the addition of uniform roll at every base-pair step tilts the base-pairs with respect to an overall helix axis, which is vertical in the picture.

The key to the calculation is to regard the central "wire," which is shown connecting base-pairs in models (a) and (c) of Fig. 3.14, as a "twisted ribbon." Since the blocks are attached locally perpendicular to the wire, tilt = $90° - \alpha$.

In this application, t becomes the local twist T (°/bp) between successive base-pairs, while k becomes the roll angle R (°/bp). N^* is the number of steps per complete helical turn; and so $360°/N^*$ corresponds to the global twist T_G (°/bp) measured with respect to an overall helix axis, i.e. an imaginary vertical line in Fig. 3.14(c). Formulas from p. 99 may thus be re-written in the present context:

$$\tan \alpha = T/R, \text{ so } \tan(\text{tilt}) = R/T; \ T_G^2 = T^2 + R^2.$$

a Given the T_G and tilt values shown below for idealized "A," "B," and "C" forms of DNA from fibers (see Chapter 8), use the formulas above to obtain the local twist T and the roll R in each case.

Fiber model	T_G (°/step)	Tilt (°)
"A"	32.7	+20
"B"	36.0	0
"C"	40.0	−10

b Inspection of Fig. 3.14(c) shows that it has global twist T_G = 36°. Use the formulas above to obtain the local twist T, for roll R = 12°. (Note that in Chapter 3 it was stated that $T \approx 36°$ for this model: thus the slight difference between values of global twist T_G and local twist T was overlooked there.)

5.3 a In Exercise 4.6 you were asked to compute the quantity which is now described on p. 104 as the "second sum," for five different repeating sequences R_n of roll angles. Confirm by direct calculation (suggestion: extend the table of Exercise 4.6 to five columns) that the "first sum" is precisely zero in all of these five cases. For ease of calculation, we fixed the roll angles in Exercise 4.6 always to give a first sum of zero; but this is not generally the case.

b Perform a calculation as set out on p. 104 for the case

$$R_n = (0°, 10°, 10°, 10°, 0°, 0°, 0°, 0°, 0°, 0°)$$

in order to find the curvature k and its phase. Is the first sum equal to zero in this example? Which step has the maximum roll R?

Note on the computation of $\arctan(f/s)$, where f = first sum and s = second sum. Suppose, for example, $f = 1.2$, $s = -1.6$. Then $f/s = -0.75$. If you ask your calculator for $\arctan(-0.75)$ it will probably give $-36.9°$, although another equally valid answer is $+143.1°$; for there are always two angles in the 360° circle, differing by 180°, whose tangents are identical. In the present case we must choose only one of these two angles. The rule which we need here is that if θ is the correct answer, $\sin \theta$ and $\cos \theta$ have the same signs as f and s, respectively. Here, $\sin -36.9° = -0.6$ and $\cos -36.9° = +0.8$, while $\sin 143.1° = +0.6$ and $\cos 143.1° = -0.8$; and so it is the second answer, 143.1°, which is correct.

5.4 Consider a long DNA molecule having a repeating sequence of the kind $A_4N_6A_4N_6A_4N_6 \ldots$, or $(A_4N_6)_m$, i.e. a 10-base-pair repeat of the kind $A_4 N_6$, where N stands for any base other than A.

a Taking $R = 0°$ for steps AA, and $R = 3.3°$ for all other steps, convert this sequence repeat into the corresponding roll angle repeat R_n for $n = 0$ to 9, where $n = 0$ corresponds to the first AA step in the repeat. Compute the first and second sums as on p. 104, and hence evaluate k; and finally express this in units of °/bp. Which step has the largest value of roll R in the Fourier wave?

b Taking twist values $T = 35°$ for step AA, and $T = 34°$ for all other steps, calculate the overall twist t for a 10-step repeat (i.e. sum the T values and then subtract 360°); and express this also in units of °/bp.

c Using the values of k and t as calculated above, find the radius r of the superhelical curve which is made by this repeated-sequence DNA, and the number N^* of base-pair steps in a complete superhelical turn. Use the formulas on p. 99; and note that r is given here in bp units: $1 \text{ bp} \approx 3.3\text{Å}$.

5.5 Repeat Exercise 5.4, but using instead sequence $(A_6N_4)_m$.

5.6 Repeat Exercise 5.4, but using the sequence $(A_6N_5)_m$. Here, in working out k it will be necessary to use multiples of $360°/11$ $= 32.7°$ for $n = 0$ to $n = 10$; i.e. first sum $= \sum_{n=0,10} R_n \sin(32.7°n)$, etc. The value of k will be in degrees per 11-step repeat; but this should be expressed in °/bp for use in the formulas on p. 99.

   Which of the sequences $(A_4N_6)_m$, $(A_6N_4)_m$ and $(A_6N_5)_m$ form left-handed superhelices, and which form right-handed superhelices?

CHAPTER 6

DNA Supercoiling

In Chapter 4 we explained how a DNA molecule must twist—or untwist—and curve, in order to carry out its various functions in biology. For example, DNA has to untwist near the start-sites of all genes, often at TATA sequences, so that RNA polymerase can construct new RNA strands according to the rules of Watson–Crick base-pairing. Similarly, DNA must untwist at all origins of replication, so that DNA polymerase can construct a new DNA strand in readiness for cell division. Lastly, DNA must curve around a variety of proteins in the cell. Some of these proteins help to package the DNA into a compact form, while others help to control the activity of some particular gene.

In Chapter 5 we went on to explain how DNA can twist and curve at the same time, so as to form long, regular supercoils or spirals. Some DNA molecules are intrinsically twisted and curved, on account of their base sequences, but others become twisted and curved only when they bind to certain proteins.

In the present chapter we shall take the subject of supercoiling one step further, by describing DNA molecules which are not curved either on account of their base sequences, or as a consequence of binding to proteins, but which coil through space nevertheless on account of *torsional stress*. And in order to provide a simple introduction to what is admittedly a rather difficult subject, let us first see how torsional stress can affect the shapes of some familiar household objects.

Figure 6.1(a) shows a piece of telephone cord. This sort of cord has a regular coiled shape because it has been manufactured with inbuilt curvature and twist. It coils through space naturally, like the intrinsically curved DNA molecules described in Chapter 5. Figures 6.1(b) and (c), on the other hand, show a piece of ordinary electric power cord. This kind of cord is manufactured without any inbuilt curvature or twist, and so it generally prefers to lie straight across a floor or table, or else in the kind of gently curving, broad loop sketched in Fig. 6.1(b). But if you take one

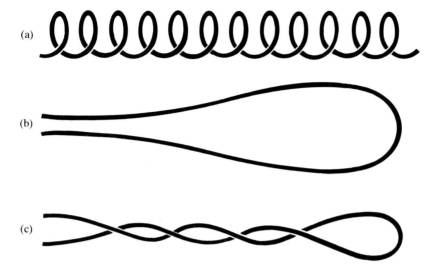

Figure 6.1 Everyday models for the supercoiling of DNA. In (a), a telephone cord coils through space naturally because it has inbuilt curvature and twist. In (b), an electric power cord has no inbuilt curvature or twist; but if you wind a few turns of twist into it, as in (c), it will cross over itself approximately once for every 360° turn of twist that you introduce.

end of this electric power cord in each hand, and wind a few turns of twist into it, then it will adopt the kind of shape shown in Fig. 6.1(c): the cord will cross over itself roughly once for every 360° turn of twist which you introduce. If you then release either end of the cord from the constraints of your hand, and shake it a bit, the free end will rotate in a reverse sense to eliminate the added twist; and the cord will return again to the uncrossed shape of Fig. 6.1(b).

Most DNA molecules, even though they are not intrinsically curved, can coil though space in the manner of the electric power cord shown in Fig. 6.1(b) and (c). It is very easy for a long DNA molecule to lose or gain a few turns of twist. This can happen, for example, by its binding to certain proteins, or by its unwinding locally during the synthesis of RNA; and if the two ends of the DNA are not free to rotate, then even a small change in twist can cause the path of the DNA to coil through space. Furthermore, there are several ways by which the two ends of a DNA molecule can be prevented from rotating. For example, the two ends can be joined with one another to form a closed circle of DNA, as shown in Fig. 6.2(a) and (b). Here, the ends of our telephone cord and electric cord have been joined into shapes which are complex and yet "circular,"

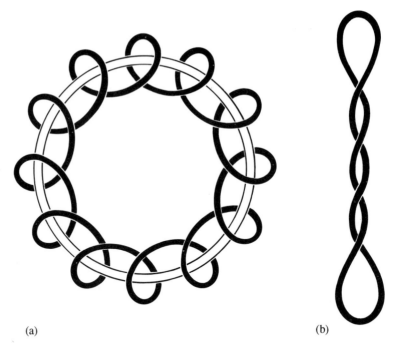

(a)   (b)

Figure 6.2 Two general varieties of DNA supercoil. In (a), the DNA coils into
a series of spirals about an imaginary toroid or ring (shown here by open
lines), and so this kind of wrapping is known as "toroidal." In (b), the DNA
crosses over and under itself repeatedly, and so this kind of wrapping is
known as "interwound."

in the sense that they are endless. There are many examples of
circular DNA in living cells.

Another example of end-restraint is shown in Fig. 6.3. Here,
a long piece of linear DNA has been divided into a series of
loops, and the two ends of each loop are constrained from
rotation where they attach to some sort of supporting structure
at the base of every loop. This kind of looped-linear arrangement
is thought to be typical of the chromosomal DNA molecules
which are found in higher organisms. We say more about this
in Chapter 7. Both kinds of DNA, the circular and the looped-
linear, will form supercoils upon any internal change of twist,
because their ends are not free to rotate. In fact, there exist in
our cells a variety of enzymes called "topoisomerases," as
mentioned in Chapter 4, that can cut the DNA temporarily so
as to make a free end, and thereby relieve some of the effects
of supercoiling. Without such enzymes, the DNA would get
tangled about itself during normal cell function, and so it could
hardly act as the genetic material. A full discussion of these
highly complex enzymes is beyond the scope of our book,

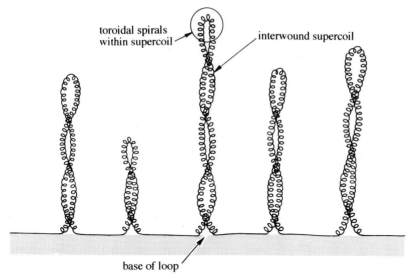

toroidal spirals
within supercoil

interwound supercoil

base of loop

Figure 6.3 The division of a long, linear DNA molecule into loops generates end-restraint at the base of every loop, if the two ends are attached to some support or "scaffold." This kind of looped-linear arrangement is thought to be typical of the chromosomal DNA found in higher organisms.

however; so the reader should consult some of the references listed at the end of this chapter to learn more about them.

In any case, once we know that the ends of a DNA molecule are fixed, then we can identify two different kinds of supercoiling, which are epitomized by the two shapes shown in Fig. 6.2(a) and (b). The circular DNA in (a) consists of a series of open spirals that wind around an imaginary ring, or toroid;[1] this kind of supercoiling is known as "toroidal." The circular DNA in (b), in contrast, winds above and below itself several times, and this kind of supercoiling is called "interwound." In practice, real DNA supercoils may contain portions of both the toroidal and interwound geometries. Thus, where certain parts of the DNA are highly curved, on account of either the base sequence or due to wrapping around a protein, one may find toroidal structures, since the DNA in a toroidal supercoil is highly curved throughout. Alternatively, if such curved portions of the DNA are not very long, they may locate themselves at the two strongly curved end-loops of an interwound supercoil, as shown at the top and bottom in Fig. 6.2(b). Sometimes the interwound and toroidal geometries may occur together, as in the looped-linear DNA which is shown schematically in Fig. 6.3. On a small scale, within any loop, the coiling is toroidal on account of the wrapping of DNA around protein spools; but on a large scale, over the full length of any loop, the structure is interwound. You often see this kind of arrangement in telephone cords, if people habitually rotate the handset.

In general, supercoiled DNA has the shapes seen in Fig. 6.2 because it either has more turns of twist, or fewer turns of twist, than the underlying, relaxed, right-handed double helix from which it is made. DNA with more than the natural number of turns is known as *overwound*, while DNA with less than the natural number of turns is known as *underwound*.

Is there any way in which we can tell, by looking at pictures such as those shown in Fig. 6.2, whether the supercoiled DNA is overwound or underwound? And can we say by how many turns the DNA is overwound or underwound? Oddly enough, the second question is easier to answer than the first. Recall that we said above, in relation to the electric power cord of Fig. 6.1(b) and (c), that the cord crosses over itself approximately once for every 360° turn of twist which is wound into it. This idea is rather general, and it may be applied also to both of the supercoils of Fig. 6.2. Thus the supercoil (b), which contains four crossovers, is either underwound or overwound by about four turns, while supercoil (a) contains 12 crossovers, and is underwound or overwound by about 12 turns. Yet it is not so easy to decide whether these supercoils are overwound or underwound; before we can say anything on that point, we

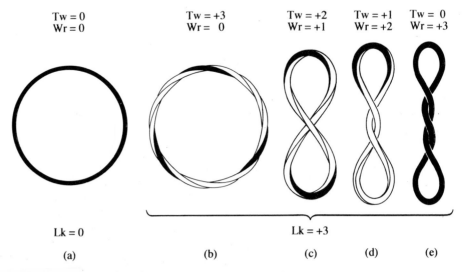

Five closely related circular DNA molecules: (a) and (b) show open circles, while (c), (d) and (e) show interwound supercoils. The DNA in its stress-free, relaxed form is drawn as a rubber rod of square cross-section, with one face painted black.

must learn more about the theory of supercoiling, as described below.

When circular DNA molecules are isolated from the cells of our bodies, or from bacteria, they are generally found to lack one turn of twist for every 17 turns of stable, right-handed double helix. Thus, they are said to be underwound, or negatively supercoiled, by about 6%. DNA molecules which contain *extra* turns are not found in Nature, except under special circumstances; we shall explain below why this is the case.

It might seem unlikely that one could ever arrive at a single scheme of explanation that would account for the shapes of all possible varieties of DNA supercoil, whether toroidal or interwound. Nevertheless, such a scheme has been invented by the mathematicians James White and Brock Fuller. It is based on a branch of mathematics known as "topology," which concerns itself with how things change shape when they go from one form to another. That is just what we want here, because DNA supercoils have lots of different shapes even though they are all made from the same kind of right-handed double helix.

We shall explain the application of topology to DNA supercoiling in two steps. First we shall deal with interwound structures, and then with toroidal ones. You will see when we have finished that we have actually used the same scheme in both cases.

Figure 6.4 shows a series of five, closely related, circular[1] DNA molecules. Two are in the form of open circles or simple rings, while three are in the form of interwound supercoils. In all cases, the DNA has been drawn as if its relaxed, stress-free form were a long rubber rod of square cross-section, with one face painted black. For example, the open circle in Fig. 6.4(a) is entirely black on one side. This means that it has exactly the same twist as relaxed, linear DNA of the same length; it is neither underwound nor overwound. Above each molecule in Fig. 6.4 we have added two symbols: Tw for "twist" and Wr for "writhe." To a first approximation, Tw tells by how many turns the rubber rod twists as it goes once around the circle, and Wr tells how many times the rod crosses over itself within any molecule. Thus, Tw = 0 for the open circle in Fig. 6.4(a), because there we can see the black face of the rod everywhere; and Wr = 0 because the rod does not cross over itself.

Figure 6.4(b) shows an open circle with three extra turns of twist. This twisted configuration can be obtained from the open circle of Fig. 6.4(a) by cutting the rod in one place, inserting there three full turns of twist and then closing it up again. It is easy to confirm that the open circle (b) contains three extra turns of twist, because the black face of the rod now shows at three locations as you go around the circle. You can also see that this new twist is right-handed, because the rod rotates in a clockwise sense, like a corkscrew, as you go around the circle. Accordingly, open circle (b) has been labeled as $Tw = +3$: the positive sign indicates that the extra twist is right-handed, i.e. in the same sense as that of the underlying DNA double helix. $Wr = 0$ again here, of course, since the rod in Fig. 6.4(b) does not cross over itself.

Now, if you make a model of the twisted circle (b) from a rubber rod, or from a long leather belt, you will find right away that this shape is unstable. Thus, if you hold the twisted ring firmly down onto a table it will look like (b), but if you let go it will collapse quickly into one of the other forms, usually into (d) or (e).

In each of the forms (c), (d) and (e), the DNA crosses over itself at least once, and the value of Wr shown above each drawing records the number of crossovers: 1 for (c), 2 for (d), 3 for (e). Note that for each new crossover, the circle loses one turn of twist. Thus, Tw decreases from $+3$ in (b), to $+2$ in (c), to $+1$ in (d), and finally to 0 in (e). You can confirm the value of Tw recorded in each drawing by counting the number of times the black face appears and disappears as one goes around the circle. It is this decrease in Tw which provides the driving force for the rod to collapse spontaneously from shape (b) to (e). For most rubber rods, or leather belts, or DNA molecules, it is easier for the rod to cross over itself repeatedly than to alter its twist: the rod responds to the overwinding by *writhing* rather than by *twisting*. One could imagine, perhaps, that a very special rod could be constructed that would behave in the opposite way, preferring instead to remain in shape (b) rather than going to shape (e). Such a rod would be easy to twist, but hard to bend. However, most real rods and DNA molecules prefer configuration (e) to (b).

You may have noticed that the sum of Tw and Wr remains constant at $+3$ for all of shapes (b), (c), (d) and (e) in Fig. 6.4. This is no accident; and indeed it exemplifies a result of general significance. It can be shown to hold true for supercoiled DNA of any shape or size, provided that the meanings of Tw and Wr are defined rigorously (see below), and that the DNA double helix remains uncut and intact on both strands. For example, the sum $(Tw + Wr)$ changed from 0 to $+3$ when we cut open the circle (a) and added three turns of twist; but then it did not change after that, in any of (b), (c), (d), or (e), because we did not cut the circle again.

Because the sum of Tw and Wr does not depend on the exact shape of the circle, but only on the intactness of its two DNA strands, mathematicians have given this quantity a special symbol, Lk, meaning "linking number":

Tw + Wr = Lk

Lk is known as a linking number because it is closely related to the number of times that the two sugar–phosphate chains of DNA wrap around, or are "linked with," one another. In this chapter, we have taken DNA in its relaxed state as the reference point for counting Lk. Thus Lk = +3 tells us that the DNA has three *more* double-helical turns than it would have in a relaxed, open-circular form. In general, Lk measures the total excess or deficit of double-helical turns in the molecule. So, when we say that a DNA molecule is "overwound by three turns" we mean, precisely, "Lk = +3." Note, in particular, that Lk can only be an integer, because the DNA can only join to itself by some integral number of turns.

Do Tw and Wr have to be integers also, or can they be real numbers such as +0.5 or +2.5? All of the pictures in Fig. 6.4 show values of Tw and Wr that are integral, but this is not necessarily so in all cases. For example, one could make an interwound supercoil with Tw = +0.5, Wr = +2.5 by rotating the lower lobe of supercoil (e) through an angle of 90° relative to the upper lobe. Then it would lie midway in shape between (d) and (e), which is satisfactory physically but hard to draw. A further untwisting of the lower lobe by an additional 90° would convert Fig. 6.4(e) all the way to (d), and yield Tw = +1.0, Wr = +2.0. So we see that neither Tw nor Wr needs to be an integer.

We have reached a point where we need more precise definitions of twist and writhe than those which we have been using so far. Let us consider Tw first. In Chapter 5 we introduced a special twist t, measured in units of degrees per base-pair, to account for the long-range coiling of DNA through space. Thus $t = 0$ for a plane curve, but t is negative for a left-handed spiral, or positive for a right-handed spiral. The twist Tw as described here is very closely related to the twist t of Chapter 5. To find the value of Tw, we can simply take the sum of t values for all of the base-pair steps in the circular DNA, and then divide the total by 360°, in order to express the result in units of helical turns. Often we do not have to use this complex definition if we wish to find the value of Tw. For example, we can evaluate Tw by inspection for each of the drawings in Fig. 6.4. But for more complicated shapes, in which the value of Tw is not an integer, it would be necessary to do the calculation in full.

Next let us consider *writhe*, which we have thought of in terms of the number of times the rod crosses over itself. The crucial point about Wr is that it is a measure of the shape of the DNA as a three-dimensional curve through space. Previously we counted the number of crossovers of the DNA in a single view in order to estimate Wr. All we need to do to get Wr accurately is to count the number of crossovers that can be seen in many different randomly chosen views of the structure, and then take the average of all of these to get the actual value of Wr. This is not a hard concept to grasp, if we think of taking a large number of snapshots of the DNA as it tumbles randomly through space, due to thermal motion. In practice, however, this may not be such a straightforward procedure, for in some views there can be many crossovers, some of which will cancel each other out: see Exercise 6.3. We shall avoid such difficult ideas in the main part of this chapter, so that only the specialist need worry about them.

Now the diagrams in Fig. 6.4 that we have been studying so far were drawn for positive Lk, i.e. for overwound DNA. You may recall that DNA in living cells is normally not overwound, but rather underwound, and so its value of Lk is negative. Therefore, in Fig. 6.5, we have provided a corresponding set of pictures for negatively supercoiled, underwound DNA. You can see by comparing the two figures that the twist is now left-handed; for example, Tw = -3 and counterclockwise in Fig. 6.5(b). Also, the way in which the DNA crosses over itself (or Wr) is subtly different in Figs 6.4 and 6.5. Thus, the DNA crosses over itself in a left-handed fashion in Fig. 6.4(e) to give Wr = $+3$, but in a right-handed fashion in Fig. 6.5(e) to give Wr = -3. In fact, the handedness of the crossovers in any interwound supercoil enables you to say definitely whether the DNA is underwound or overwound, simply by looking at a picture. For open circular DNA, where there are no crossovers, you have to look at the twist of the rod to see whether the DNA is overwound (clockwise twist) or underwound (counterclockwise twist). The drawings in Figs 6.4 and 6.5 should serve as reliable references for further study.

At the beginning of this chapter we described two general classes of supercoil, known as interwound and toroidal. So far we have investigated only the interwound form. But now that we are familiar with the meanings of Lk, Tw and Wr, it should be a relatively easy matter to analyze the shape of a toroidal supercoil, such as that shown in Fig. 6.2(a).

The best way of proceeding might be to examine in detail one small portion of a toroidal supercoil, say a segment of two superhelical turns. Figure 6.6(a) and (b) show how two such toroidal, superhelical turns can be generated by the collapse of a highly twisted piece of DNA. In diagram (a) we have drawn a twisted ribbon, with one of its edges shaded. It represents a piece of DNA which has been underwound by two turns. The ribbon is attached at either end to a block, so that its two ends cannot rotate: this device enables us to consider a small piece of circular DNA in isolation. For this DNA you can see that Tw = −2 and Lk = −2, since Wr = 0 (there are no crossovers in any perspective).

In Fig. 6.6(b) we have moved the two blocks closer together, so that the twisted ribbon can collapse into part of a toroidal coil, like the coil made by a snake as it lies in the grass. The ribbon now makes two flat, left-handed turns: you can verify that they are left-handed by doing a simple experiment with a short piece of a leather belt, or a strip of paper. In this configuration both coils are almost planar, so Tw ≈ 0, and the linking number has not changed from its previous value of −2. Thus we find that Wr = Lk − Tw = −2. The main point which emerges from this exercise is that two flat, left-handed toroidal

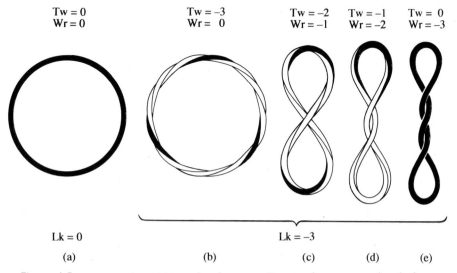

| Tw = 0 | Tw = −3 | Tw = −2 | Tw = −1 | Tw = 0 |
| Wr = 0 | Wr = 0 | Wr = −1 | Wr = −2 | Wr = −3 |

Lk = 0   Lk = −3

(a)   (b)   (c)   (d)   (e)

Figure 6.5 Five circular DNA molecules as in Fig. 6.4, but now with a linking number Lk of the opposite sense. These DNA molecules are underwound, with Lk negative, while the ones in Fig. 6.4 were overwound, with Lk positive.

Tw =–2, Wr = 0

(a)

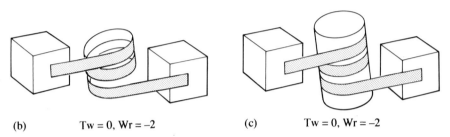

(b)   Tw = 0, Wr = –2   (c)   Tw = 0, Wr = –2

Figure 6.6 A highly twisted ribbon will collapse spontaneously into part of a toroidal supercoil. In (a), the two ends of the ribbon are held apart by their attachment to blocks, so that Tw = −2. In (b), the blocks move together so that the ribbon can collapse to Wr = −2. In (c), a cork or protein spool stabilizes the shape of the ribbon shown in (b).

coils have a writhing number of −2. Thus, in going from Fig. 6.6(a) to (b), the twist of the DNA has been transformed into writhe.

Let us now return to the toroidal circle of Fig. 6.2(a). Its coils are certainly left-handed, but they are not so flat as those of Fig. 6.6(b). If we imagine a simple transformation by which the entire toroid is pushed inwards, so that its radius becomes smaller, then the coils will become flatter. In such a case the writhing number can be assigned as −2 for each pair of coils, just as in the example above; and so Wr = −12 for the entire molecule. On the other hand, if the supercoil were to be pulled out to a much larger radius, so that it became a simple, open circle, then it should have Wr = 0, or Tw = −12. In the configuration shown, Wr will lie somewhere between these two extremes, and closer to −12 than to 0; so perhaps Wr = −10.

You may have noticed in our pictures that a *left-handed* coil in the toroidal form of Fig. 6.2(a) gives negative writhe, while a *right-handed* coil in the interwound form of Fig. 6.2(b) also gives negative writhe. Is there some sort of mistake here? Surely the rule of crossovers should be consistent among different types of supercoil, such as interwound and toroidal? In fact, there is no mistake at all. The two sets of pictures have been derived unambiguously from physical experiments with underwound rubber rods or leather belts. The key point here is that we need a way of allocating a *sign* to any given crossover which we can see, in order to say unambiguously whether it contributes positively or negatively to writhe. This is not a trivial exercise, and so we have consigned it to Exercise 6.3.

We have now described two different kinds of supercoiling for DNA—toroidal and interwound. But what are the relative stabilities of these two forms? In other words, when will a DNA molecule be interwound, and when will it be toroidal? The interwound shape is usually very stable, and most underwound or overwound DNA molecules will naturally adopt an interwound shape, in the absence of other forces. But the proteins that associate with DNA in living cells can sometimes change the situation dramatically, and favor the toroidal over the interwound form by wrapping the DNA around themselves.

For example, consider the cork which has been inserted between the two turns of ribbon shown in Fig. 6.6(c). This cork represents a typical protein "spool" around which the DNA can wrap, and around which it does wrap in a left-handed sense in the chromosomes of most forms of life on Earth. If the DNA or ribbon in Fig. 6.6(c) were to be cut free from the two blocks at either end, it would stay wrapped around the "sticky" protein spool; whereas if it were cut free in the absence of a spool, as in Fig. 6.6(b), it would immediately spring back into a straight configuration. When we isolate DNA in the laboratory in pure form from any kind of cell or cells, at some point in the procedure we must strip off the proteins around which the DNA was originally wrapped, without breaking either of its two double-helical strands. In other words, we must remove the cork from the arrangement shown in Fig. 6.6(c), without cutting the DNA free from either of its two end-blocks. Naturally the "naked" DNA will first spring out to the highly twisted form shown in Fig. 6.6(a), and then it can collapse into an interwound supercoil as shown in Fig. 6.5(e), because it has lost the curvature which stabilized the toroidal form. Therefore, we can expect to see highly interwound supercoils in the preparations of pure DNA which we make from living cells, after removal of various proteins. Incidentally, this is why DNA supercoils in Nature are always underwound rather than overwound: the DNA always coils around proteins in the cell nucleus in the form of a left-handed toroidal spiral, giving negative Lk.

Some typical preparations of purified, protein-free DNA from bacteria are shown in Fig. 6.7(a), exactly as they appear in the electron microscope. These DNA molecules are circular forms of length 7000 base-pairs, and they are underwound on the average by 40 turns, i.e. Lk = −40. A linear, relaxed double helix of the same length would contain 7000/10.6 = 660 turns of right-handed DNA, given a typical helical twist of 10.6 base-pairs per turn; so these DNA molecules contain just 620 turns and are underwound by 40/660 = 6%.

Note that the supercoils shown in Fig. 6.7(a) have a branched formation of the kind indicated schematically in Fig. 6.7(b). This does not in itself have much effect on either the topology or the energetics of the structure. Indeed, the branched structure could "migrate" smoothly into an unbranched form that would be much larger overall, if conditions were right.

Assuming that these DNA molecules are all of the interwound form, can we determine their Lk, Wr and Tw? A value for the linking number of Lk = −40 can be determined independently of the electron-microscope pictures, by studying the mobility of such DNA during gel electrophoresis: this method will be described in Chapter 8. An approximate value for the writhing number, Wr = −36, can be determined by counting the mean number of crossovers per DNA molecule, as seen in many pictures from electron microscopy such as those of Fig. 6.7(a). Finally, by subtraction, Tw = Lk − Wr = −4. Thus, the preferred interwound structure is somewhat similar to the idealized shape shown in Fig. 6.5(e), since Wr = 0.9 Lk, and Tw = 0.1 Lk. In other words, the DNA which has been underwound finds it more favorable energetically to cross over itself repeatedly, than to alter its twist.

Some typical pictures of toroidal supercoils, where the coiling of the DNA has been stabilized by wrapping about proteins, will be shown in Chapter 7.

That concludes our survey of the physical properties of supercoiled DNA. We can now concern ourselves with its biological properties. Supercoiling is important in biology because it helps the DNA to unwind, so as to promote the synthesis of new RNA or DNA strands. As shown in Fig. 6.8, the various polymerase proteins that copy pre-existing DNA into new RNA or DNA must first unwind the DNA by one to two turns in the locations where they wish to act. This then enables them to "read" the unpaired bases on one strand of the DNA, so that they can assemble a new strand according to the rules of Watson–Crick base-pairing. Sometimes the polymerase

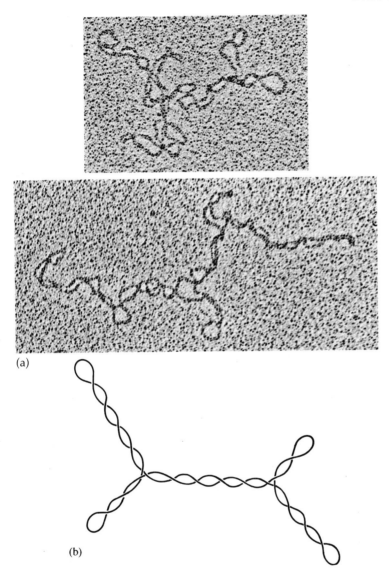

(a)

(b)

Figure 6.7 (a) Electron micrographs of negatively supercoiled, interwound DNA as prepared in pure form from *E. coli* bacteria. Each DNA plasmid or ring is 7000 base-pairs long, and has a mean Lk = −40. Courtesy of Christian Boles, Nick Cozzarelli, and James White; and from *Journal of Molecular Biology* (1990) **213**, 931–51. (b) Branched path of the interwound DNA shown in (a), in schematic form. Such branching has little effect on the parameters Lk, Tw, and Wr.

itself does not unwind the DNA, but rather its "helper" proteins. In any case, the DNA must be unwound or else accurate synthesis cannot proceed.

We said above that most of the missing turns in any negatively supercoiled DNA molecule are stored in the form of writhe Wr,

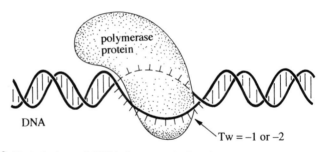

Figure 6.8 Untwisting of DNA by a typical polymerase protein, here shown schematically. Usually the twist of the DNA must be reduced by one or two turns in order for an RNA polymerase molecule to function.

whether by crossovers in an interwound supercoil, or by flat spirals in a toroidal supercoil. How, then, can supercoiling produce a reduction of twist Tw by one or two turns, as is needed for a polymerase protein to unwind DNA in various locations? Clearly, the DNA must be able to vibrate or fluctuate in solution, as a kind of Brownian motion, from shapes with high writhe to shapes with high twist. For example, an interwound supercoil might vibrate from the shape shown in Fig. 6.5(e) to any of the shapes shown in Fig. 6.5(d), (c) or (b), in order to generate twist. Similarly, a toroidal supercoil might vibrate from the shape shown in Fig. 6.6(b) to that shown in Fig. 6.6(a).

Unfortunately, we have no direct experimental evidence today as to how a DNA molecule fluctuates in solution over a large scale. We know, through probing for single-stranded regions using enzymes and chemicals, that negatively supercoiled DNA vibrates much more efficiently than relaxed DNA to yield negative Tw; and we know also that many genes require negative supercoiling in order to be transcribed by RNA polymerase; but we do not know how DNA changes its shape over a large scale, to produce vibrations that lead to the generation of twist. All we have today are a great many lines of indirect evidence to suggest what might be going on. Furthermore, our indirect data are limited to observations about bacterial genes, because the genes in higher organisms are so poorly understood that one cannot draw any firm conclusions about how they work.

First, we have the observation that many different bacterial genes will synthesize more RNA, if the DNA in their "promoter" regions is appropriately curved due to its base sequence, than if this DNA is inappropriately curved or straight. The structure of a typical bacterial gene is shown in Fig. 6.9(a). There we can see that an RNA polymerase protein starts to make RNA, or "messenger RNA," just upstream of the long segment of DNA that codes for protein. The RNA copy of DNA then travels from the bacterial chromosome to the protein-making machinery, or ribosome, where the series of bases A, U, C, and G along its length specifies the synthesis of some protein with a certain number and ordering of amino acids. More protein is synthesized if there are more RNA molecules of a certain kind, and the number of RNA molecules depends in turn on how often RNA polymerase can initiate the synthesis of new RNA chains at any promoter.

Why should the curvature of DNA within the promoter affect how often RNA polymerase will start to make RNA? It makes no sense, if we consider the problem just in *one dimension* as shown in Fig. 6.9(a). It begins to make sense in *two dimensions*, when we consider that the curved DNA might wrap around the polymerase protein, as shown in Fig. 6.9(b), and so help it to bind at the promoter. In fact, there is considerable physical evidence, especially from the work of Andrew Travers, Henri Buc and others, that DNA wraps by as much as 180° around *E. coli* RNA polymerase before it starts to make RNA chains (the exact degree of curvature is not certain).[2] Furthermore, it is well-known that this RNA polymerase not only docks onto the curved DNA, but also probes for the identities of bases in two regions of the promoter called "−35" and "−10," as indicated in Fig. 6.9(b). So the DNA cannot curve just in any arbitrary direction in order to help RNA polymerase bind at the promoter; rather, it must curve in an appropriate direction, so as to locate the sites of contact between RNA polymerase and the −35, −10 regions along the inside of the curve.

The problem makes even more sense when we consider the possible wrapping of DNA about the polymerase in *three dimensions*. As shown on the left-hand side of Fig. 6.9(c), DNA can wrap into one-half turn of a left-handed, toroidal supercoil about the polymerase protein, yielding Wr ≈ −0.5 turn. Due to Brownian motion in solution, this DNA will vibrate fairly often into an extended form with Tw ≈ −0.5 turn, as shown on the right-hand side of the figure (the protein being no doubt flexible enough to permit such motion). Thus, the wrapping of DNA about RNA polymerase could conceivably help the polymerase to unwind DNA near the start of the gene, i.e. the −10 region, by means of thermal vibration in solution. Indeed, the sequence of the DNA in the −10 region is usually of a kind TATA, that

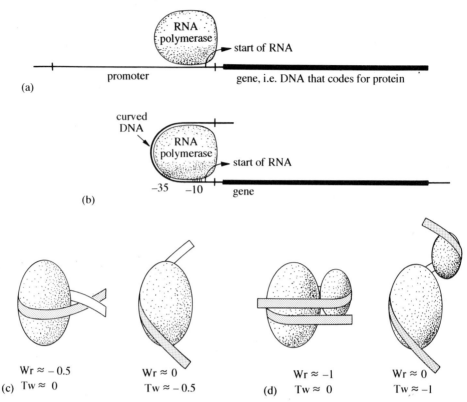

Figure 6.9 Highly schematic pictures of the structure of a typical bacterial gene, and its interaction with RNA polymerase. In (a), the polymerase protein binds just upstream of the DNA that codes for protein, and within the region of the "promoter." In (b), the presence of curved DNA within a promoter helps the polymerase to bind more tightly to the DNA, and hence to synthesize more RNA chains. Specific contacts between the protein and the DNA cluster in two regions that lie 35 and 10 base-pairs, respectively, upstream of the start-point for making RNA. The DNA unwinds at the −10 region (often of sequence TATA) to let the polymerase function as in Fig. 6.8. In (c), the DNA wraps into one-half turn of left-handed, toroidal supercoil about the polymerase, so that it can generate −0.5 turn of twist by thermal vibration in solution. In (d), an "activator" protein wraps the DNA by one full turn about the polymerase to yield −1.0 turn of twist as the DNA vibrates in solution.

helps the DNA to unwind locally in response to long-range torsional stress.

Of course, these fluctuations of the supercoiled DNA cannot select very precisely for which part of the DNA is to be unwound, because the torsional stress may extend over many helix turns. The easy unwinding of a TATA sequence in the −10 region is one factor that selects for a precise site of unwinding, while another might be the contacts of the polymerase protein with base-pairs in the −35 region. In any case, it should be emphasized that the protein itself as shown in Fig. 6.9(c) might be expected to change its shape, in order to follow the path of the fluctuating DNA; but presently we know little about how this is accomplished.

Curved DNA is also found where DNA has to unwind to initiate the synthesis of DNA, at origins of replication in bacteria and in certain viruses. The DNA there wraps into a long, left-handed spiral about a series of small proteins that seem flexible enough to let the DNA vibrate like a bed-spring. When you mix these proteins with their preferred origins of replication on pure, negatively supercoiled DNA, they cause the DNA to unwind through thermal vibration at sequences near one end of the spiral, in a way that can be detected through the use of enzymes and chemicals that react specifically with single-stranded DNA.

So a large body of evidence seems to suggest that DNA can vibrate from Wr to Tw, in order to help DNA unwind for the synthesis of new RNA or DNA chains. These fluctuations will occur more readily in negatively supercoiled than in relaxed DNA, thereby accounting for the fact that negative supercoiling can activate genes.

There is one final point that we must consider. In any such Wr-to-Tw vibration, the size of Wr sets a limit on the amount of Tw that can be obtained. For example, a Wr of −0.5 turn can yield only a Tw of −0.5, but a Wr of −1 can yield a larger Tw of −1, close to that required by RNA polymerase (see Fig. 6.8). Relevant to this point, there are known to be a wide variety of proteins called "activators" in bacteria, that seem to make genes work better in the following way: they bind to the DNA upstream of the polymerase, and curve it by an extra 180° as shown in Fig. 6.9(d). It used to be thought that these activator proteins might simply stick to the polymerase, thereby helping it to bind the promoter. While this may still be true in part, more recent data have shown three things. First, the activator proteins do not have to occupy any precise location along the DNA relative to the polymerase, so long as they curve the DNA in a correct direction. Second, some proteins that curve the DNA can act as "repressors" of gene activity in one location, by binding competitively to the same piece of DNA as that preferred by RNA polymerase; then they can act as "activators" of gene activity, when moved to a new location upstream of the polymerase as in Fig. 6.9(d). Third, protein-free DNA can provide for a similar activation of gene activity, if this DNA is appropriately curved on account of its base sequence. So we see that these activator proteins (and the curved DNA) may act simply by increasing the value of Wr, and thereby increasing the value of Tw that can be generated as the DNA vibrates in solution.

In summary, the theory of supercoiling has three strong implications for our understanding of the ways in which genes work. First, the DNA must vibrate from Wr to Tw to generate twist; second, toroidal supercoils and interwound supercoils will vibrate differently on account of their different shapes; and third, the larger the value of Wr within a given region, the more Tw that can be obtained. This would seem to be one of the most interesting but least explored subjects in molecular biology, where theory has gone past experiment. We really do not need any more circumstantial evidence on this matter from biological experiments; rather we need new kinds of direct evidence from physical or chemical experiments, to learn how supercoiled DNA vibrates near genes, perhaps in association with protein molecules.

Note

1. See Appendix 1.
2. Recent work by W. Wahli, J. Dubochet and colleagues in *EMBO Journal* **11**, 667−72 (1992) shows that *E. coli* RNA polymerase induces a strong curve of 180° in the DNA: so strong, in fact, that its binding-site on DNA often becomes the curved end-loop of an interwound supercoil.

Further Reading

Boles, T.C., White, J.H., and Cozzarelli, N.R. (1990) The structure of plectonemically supercoiled DNA. *Journal of Molecular Biology* **213**, 931–51. Electron microscope pictures of interwound supercoils as in Fig. 6.7.

Bramhill, D. and Kornberg, A. (1988) Duplex opening by *dnaA* protein at novel sequences in initiation of replication at the origin of the *E. coli* chromosome. *Cell* **52**, 743–55. The transient unwinding of DNA as induced by its wrapping about proteins, except now in the case of a bacterial origin of replication.

Collis, C.M., Molloy, P.L., Both, G.W., and Drew, H.R. (1989) Influence of the sequence-dependent flexure of DNA on transcription in *E.*

coli. Nucleic Acids Research **17**, 9447–68. The main part of the binding site for RNA polymerase can be curved DNA.

Heumann, H., Ricchetti, M., and Werel, W. (1988) DNA-dependent RNA polymerase of *E. coli* induces bending or an increased flexibility of DNA by specific complex formation. *EMBO Journal* **7**, 4379–81. A bacterial RNA polymerase curves the DNA close to where it starts making RNA.

McAllister, C.F. and Achberger, E.C. (1989) Rotational orientation of upstream curved DNA affects promoter function in *Bacillus subtilis*. *Journal of Biological Chemistry* **264**, 10451–6. Synthetic curved DNA sequences, when placed upstream of a bacterial RNA-polymerase promoter, modulate by 10-fold the amount of RNA made in a test tube or in living cells.

Millar, D.P., Robbins, R.J., and Zewail, A.H. (1982) Torsion and bending of nucleic acids studied by subnanosecond time-resolved fluorescence depolarisation of intercalated dyes. *Journal of Chemical Physics* **76**, 2080–94. The best present-day method for studying short-range DNA vibrations.

Pérez-Martín, J. and Espinosa, M. (1991) The *RepA* repressor can act as a transcriptional activator by inducing DNA bends. *EMBO Journal* **10**, 1375–82. Even a repressor protein that is designed normally to reduce the amount of RNA made, can act as an "activator" by curving DNA upstream of RNA polymerase.

Schnos, M., Zahn, K., Inman, R.B., and Blattner, F.R. (1988) Initiation protein induced helix destabilisation at the λ origin: a prepriming step in DNA replication. *Cell* **52**, 385–95. The curvature of DNA around proteins at a viral origin of replication causes transient unwinding of the DNA, if it is negatively supercoiled in a test-tube.

Schultz, S.C., Shields, G.C., and Steitz, T.A. (1991) Crystal structure of a CAP–DNA complex: the DNA is bent by 90°. *Science* **253**, 1001–7. Three-dimensional structure of a typical bacterial activator protein as bound to DNA: not only does the DNA bend sharply, but it bends into part of a left-handed toroidal supercoil.

Vinograd, J., Lebowitz, J., Radloff, R., Watson, R., and Laipis, P. (1965) The twisted circular form of polyoma viral DNA. *Proceedings of the National Academy of Sciences, USA* **53**, 1104–11. The earliest recorded work on DNA supercoiling.

Wang, J.C. (1991) DNA topoisomerases: why so many? *Journal of Biological Chemistry* **266**, 6659–62. An up-to-date survey of the actions of topoisomerases in living cells.

Bibliography

Amouyal, M. and Buc, H. (1987) Topological unwinding of strong and weak promoters by RNA polymerase: a comparison between the *lac* wild-type and UV5 sites of *E. coli. Journal of Molecular Biology* **195**, 795–808. They find a loss of linking number but no loss of twist on the initial binding of RNA polymerase to DNA; which implies that this protein induces negative writhe before it unwinds the double helix.

Bracco, L., Kotlarz, D., Kolb, A., Diekmann, S., and Buc, H. (1989) Synthetic curved DNA sequences can act as transcriptional activators in *E. coli. EMBO Journal* **8**, 4289–96. Synthetic curved DNA can replace the binding-site for the CAP protein in *E. coli*, and still activate the gene in a similar fashion.

Cozzarelli, N.R., Boles, T.C., and White, J.H. (1990) Primer on the topology and geometry of DNA supercoiling. In *DNA Topology and its Biological Effects* (eds N.R. Cozzarelli and J.C. Wang), ch. 4, pp. 139–84. Cold Spring Harbor Laboratory Press. A discussion similar to the one in Chapter 6.

Drew, H.R. and Travers, A.A. (1985) DNA bending and its relation to nucleosome positioning. *Journal of Molecular Biology* **186**, 773–90. DNA sequences from the upstream region of a bacterial promoter are shown to direct the same kind of curvature for DNA wrapped around RNA polymerase, or wrapped around the chromosomal histone proteins, or wrapped into a tight, protein-free circle.

Fuller, F.B. (1971) The writhing number of a space curve. *Proceedings of the National Academy of Sciences, USA* **68**, 815–19. Development of the concept of writhe for both interwound and toroidal supercoils.

Lamond, A.I. and Travers, A.A. (1983) Requirement for an upstream element for optimal transcription of a bacterial tRNA gene. *Nature* **305**, 248–50. DNA sequences upstream of the usual −35 and −10 regions are shown to activate a bacterial promoter 10-fold.

Exercises

6.1 For this exercise you will need a substantial length (say about 1.5 m) of flexible rod or cord. An electric extension cord, with circular cross-section about 6 mm diameter, is ideal; or else you may use rubber tubing (provided it is not *curved* when relaxed) or even a plaited rope of the kind which is sold at boat shops. For the sake of convenience, we shall refer to all of these models as "cords."

    a Use the cord to reproduce the transformation shown in going from Fig. 6.1(b) to (c). For each full turn of 360° wound into one end of the cord, how many times does it cross over itself?

    b Use the same cord to demonstrate that a left-handed toroidal supercoil, as in Fig. 6.2(a), transforms into a right-handed interwound coil, as in Fig. 6.2(b), when it is shaken out. To do this, first make a straight left-handed coil by wrapping the cord around a stick. Then carefully remove the stick and bend the coil around to make a toroid; and finally hold the two ends together in one hand, and shake it out vertically. How many times does the right-handed interwound form cross over itself, for each turn of left-handed toroid around the stick?

6.2 For this exercise you should use the same cord as in Exercise 6.1, but now you must draw a black (or coloured) line (or stripe) along the entire length of the untwisted cord, in order to signify zero twist. To make a closed loop from the cord you can fasten the two ends together with sticky tape; or you can use a short dowel if you are working with rubber tubing. Or you can hold the two ends together with your hand, as in Exercise 6.1.

Reproduce the transformations shown in Figs 6.4 and 6.5, by winding into the cord three full turns of either right-handed or left-handed twist, to give $Lk = +3$ or -3, respectively. Which of the shapes (b) to (e) is the most stable for your model? Count the number of crossovers, and then use Figs 6.4(e) and 6.5(e) as guides to decide upon the sign of Wr, by looking at the handedness of the crossovers. Also, follow the path of the black stripe in order to identify the sign and magnitude of twist Tw. Confirm that $Lk = Tw + Wr$.

Untwist the cord fully and repeat with, say, $Lk = +5$ or -5.

6.3 This is an exercise on determining the sense (+ or −) with which an individual crossover contributes towards the writhe Wr.

First, identify some particular crossover in a picture of a writhed cord. Put an arrow on the upper segment of the cord (i.e. the one nearer to you), and follow the cord round until you reach the lower segment at the same crossover. Mark an arrow on this segment toc, in the same sense as the upper arrow, when going along the contour length of the cord.

The two crossed segments should now both be marked with arrows, as shown in (a). Next, rotate this local picture of the crossover until the arrow of the upper segment points towards the top of the page. If the lower arrow ncw points from right-to-left, as in (b), then the crossover counts as +1 to Wr; but if it points from left-to-right, as in (c), then the crossover counts as −1 to Wr.

Use this double-arrow test to confirm that there is negative Wr in each of the coils shown in Fig. 6.2; and also to confirm

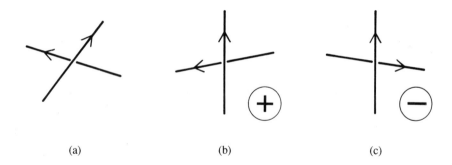

(a)   (b)   (c)

the signs of writhe Wr assigned to the coils shown in Figs 6.4(e) and 6.5(e).

(You can, equivalently, associate a "right-hand rule" with the situation shown in (b); and if you are familiar with rules of this kind in electricity and magnetism you might well find it more useful. Thus, if you straighten the thumb of your right hand and bend the fingers round, and then point the thumb along the upper segment, your fingers will indicate the direction of the lower segment, for positive Wr. But if you try this with (c), you will find that the lower segment goes the other way: so there we have a negative contribution to Wr.)

6.4 For this exercise you can use either a cord with a painted stripe, as in Exercise 6.2, or else a ribbon or strip of paper which has been coloured on one side with a felt-tipped pen, as in Fig. 6.6.

   a Wrap a portion of the cord twice, in a left-handed sense, around an ordinary soft-drink or beer can. Use your two hands to represent the two blocks to which the ends of the cord are attached, as in Fig. 6.6(c). Then remove the can from the wrapped cord, and pull the ends apart as in Fig. 6.6(a). How many turns of twist Tw have been created by the wrapping? Are these new turns right-handed or left-handed?

   b The vast majority of DNA in our chromosomes wraps about protein spools of the kind shown in Fig. 1.5. There, the DNA wraps twice in a left-handed sense around each spool, just as in our ribbon-around-the-beer-can exercise. But when the DNA is removed from such a protein spool, it turns out by experiment that only one turn of twist is created for each two turns of wrapping. In other words, Tw = −1 instead of −2 as would be expected from the simple models of part (a) and Fig. 6.6(a). By how many turns would the DNA have to be pre-twisted, before being wrapped on the spool (or else during the process of wrapping), to account for this result? Why might the protein spools be designed to lessen the expected change in twist Tw, on wrapping or unwrapping the DNA? (See Germcnd, J.E., Hirt, B., Oudet, P., Gross-Bellard, M., and Chambon, P. (1975) *Proceedings of the National Academy of Sciences, USA* **72**, 1843–47 for the first experiment of this kind ever performed.)

6.5 Suppose that DNA wraps around a cylindrical protein into n complete turns of left-handed supercoil; for example, $n = 1$ on the left-hand side of Fig. 6.9(d), or $n = 2$ in Fig. 6.6(c). If we represent the stress-free, relaxed DNA as a ribbon, we can easily find by direct experiment in such cases that $Lk = -n$, where n is a positive, whole number of turns. Thus, Lk as so defined is necessarily an integer; and the negative sign of Lk corresponds to the left-handed sense of the wrapping.

Now we can further describe the path of the coiled DNA in terms of a pitch angle α, as shown previously in Fig. 5.4. The angle α is likewise negative for left-handed wrapping. Furthermore, by combining results from Chapters 5 and 6, it can be shown that:

$$Tw = n \sin\alpha, \quad Wr = Lk - Tw = -n(1 + \sin\alpha)$$

a For $n = 1$, calculate Tw and Wr for $\alpha = 0°, -30°, -60°$, and $-90°$.

b Suppose that a hypothetical polymerase protein requires at least -0.5 turns of twist in order to bind to the DNA in the region of a TATA sequence, so that it can start RNA synthesis. If the DNA has exactly one left-handed turn of supercoiling at $\alpha \approx 0°$, as shown on the left-hand side of Fig. 6.9(d), and undergoes a transformation of the kind shown on the right-hand side, what new pitch angle α will be needed to start RNA synthesis? What would be the corresponding value of α for an initial wrapping of $n = 2$?

CHAPTER 7

The Assembly of DNA into Chromosomes

In Chapter 1 we gave a general description of the biology of a typical cell, and we explained how DNA plays a central role in this biology, by specifying the construction of protein molecules. Since then, we have considered DNA mainly as a simple, double-helical thread which undergoes transcription and replication, and which wraps itself around protein spools. Our task in the present chapter will be to describe the assembly of DNA into chromosomes. How does DNA fold into the highly compact chromosomes shown in Fig. 1.3, when a cell divides? And how does it organize itself at other stages in the life of a cell, as in Fig. 1.2, when it has to make the RNA that codes for protein? Our answers to these questions will be, unfortunately, less secure than we would prefer. Only a few aspects of chromosome structure are known with confidence, while the rest require a lot of educated guesswork.

We begin by describing in some detail the proteins which make up any histone spool, and the DNA that coils around them. Next we explain how a string of successive histone spools can coil into a so-called "300 Å fiber," if the conditions are right. Then we describe how these 300 Å fibers might fold into a series of loops along some protein "scaffold"; how genes might work along the hypothetical loops; and how the scaffold complete with loops might coil once again into the form of the compact chromosome which we can see on cell division. Finally, we explain how chromosomes are assembled from proteins and DNA, at least at a rudimentary level; and how they might be disassembled when genes are activated for the synthesis of RNA.

Now you may recall from Fig. 1.1 that the DNA in our chromosomes is compacted by a factor of about 10 000 in total, as compared to the length of a simple, double-helical thread.

When DNA wraps around the histone spools, its overall length is reduced by a factor of about 6; so when it wraps into various other kinds of structure within a chromosome, its overall length must be reduced again by a further factor of 1500. Indeed, each individual chromosome contains a remarkably long length of DNA: typically 1–10 million base-pairs in yeast, or 50–400 million base-pairs in humans. If this DNA were not highly compacted in some definite way, then life as we know it could not exist. Throughout our presentation we shall consider, in a somewhat speculative way, how this degree of compaction might be accomplished.

Let us begin with a brief history of the subject. Chromosomes were studied for many years by geneticists, but not at a molecular level. These scientists knew that chromosomes were made from a mixture of protein and DNA, but they did not know whether this mixture might possess any regular structure or organization. In 1973, Dean Hewish and Leigh Burgoyne clarified the situation dramatically: they found that the majority of the DNA in a chromosome could be digested by a DNA-cutting enzyme into many small fragments of regular size, such as 200, 400, 600, 800, etc. base-pairs.

They were able to explain their result as follows. Suppose that chromosomes are made largely from a series of nearly identical units, each consisting of certain proteins in combination with 200 base-pairs of DNA. Now if the DNA-cutting enzyme were to act at every point where it found a weakness, perhaps in the regions of DNA which lie between units, then all of the long chromosomal DNA would be cut into short pieces of size 200 base-pairs. But if the cutting enzyme were to act at random, at only a limited number of points, then this long DNA would be reduced in size just to multiple units of 200, 400, 600, 800 base-pairs, etc., simply because not every point of weakness would be cut. Furthermore, if the individual units of 200 base-pairs were not of precisely determined size, but were to vary in size, for example from 180 to 220 base-pairs, then the lengths of the multiple units would vary also: from 360 to 440, 540 to 660, 720 to 880 base-pairs, etc. In such a case, these multiple units could hardly be seen as fragments of discrete size. But in fact, multiple units of discrete size are often seen to DNA lengths as large as 2000 base-pairs in such experiments. Thus, one may conclude that certain proteins bind to this DNA in a rather precise fashion, once every 200 base-pairs, and thereby set its fundamental length.

It is now known that regular combinations of protein and DNA can be found once every 200 base-pairs along most of the length of DNA in any chromosome. These particles are called *nucleosomes*, and they are very important in biology. The fundamental length is not always exactly 200 base-pairs in every kind of animal or plant, or in every type of tissue within a particular kind of animal or plant. In fact, these fundamental "spacings" are known to vary from as short as 170 base-pairs in certain chromosomes, to as long as 260 base-pairs in others. So one suspects that there may be some variety in the kinds of protein that make up any nucleosome. We have already shown some crude pictures of nucleosomes elsewhere in the book, for example in Figs 1.5 and 6.6(c), without giving these particles a name, or explaining how they were made.

For the sake of accuracy, one should note that the repeating pattern seen by Hewish and Burgoyne in 1973 had first been seen 3 years earlier by Robert Williamson. He was studying mouse cells in tissue culture, and he saw DNA fragments of size 200, 400, 600, 800 base-pairs, etc., as products of degradation from cells that were not growing so well. But he thought that these fragments came from the incomplete synthesis of long DNA molecules, as shown by previous work, rather than from the degradation of intact chromosomes. Because of this, his work attracted little attention; yet progress in chromosome research could have proceeded more rapidly, if more people had realized its implications.

Following the correct interpretation of these data by Hewish and Burgoyne, there was a flurry of research in many laboratories across the world. For example, early in 1974 Ada and Donald Olins visualized a preparation of nucleosomes at high magnification using an electron microscope (a technique to be described in Chapter 8), and saw a series of protein "beads" along an extended DNA "string." Such a result was clearly consistent with the idea of a regular, repeating structure for DNA in chromosomes. The next significant advance came later in 1974, with a report by Roger Kornberg and Jean Thomas on the identity and approximate organization of the proteins in a typical nucleosome. They studied the physical properties of all the major chromosomal proteins, which are known as "histones"; and they found that the majority of histone types, known as H2A, H2B, H3, and H4, could associate with one another in a stable fashion to form a protein particle around which the DNA could wrap. This particle is now known as the "histone octamer," because it contains eight proteins in total: there are two copies of each of the four distinct kinds listed above. At first it was thought that all of the DNA in any 200-base-pair nucleosome might wrap solely about this histone octamer, but later it was realized that the histone octamer as such binds only to about 150 base-pairs of DNA. A ninth protein molecule, known as histone H1, was found to bind to the remaining 50 or so base-pairs. These results are summarized in a simplified way in Fig. 7.1(a), which shows schematically the various histones strung out along the DNA in one dimension.

The three-dimensional structure of the histone octamer together with its DNA was not established firmly until 1977, when John Finch, Len Lutter, Daniela Rhodes, Aaron Klug and colleagues grew crystals of the histone octamer complete with DNA, that were suitable for analysis by X-ray diffraction methods. They also used electron microscopy to study the structure at low resolution, and enzyme-digestion methods to probe the structure in solution. By piecing together data gathered from these various methods (which will all be described in Chapter 8), they were able to show that the 150 base-pairs of DNA wrap almost twice around the histone octamer into a shallow, left-handed supercoil containing about 80 base-pairs per turn. Seven years later, in 1984, the same group with Tim Richmond succeeded in obtaining the structure of the histone octamer complete with its DNA to a much higher resolution, solely by the method of X-ray diffraction. A model of their structure is shown in Fig. 7.2, where the DNA (in brown) wraps

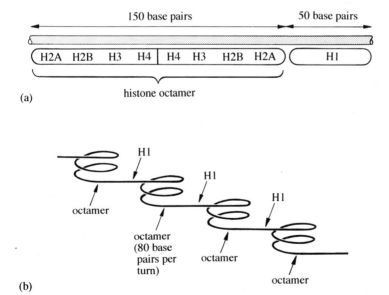

Figure 7.1 Histone proteins bound to DNA, and shown schematically in one dimension or three. The DNA is drawn as a long tube in (a), but as a long, curved string in (b).

almost twice around the outside of a "spool" made from the two copies each of histones H2A, H2B, H3, and H4, which are shown in different colors. The diameter of the protein spool is about 60 Å, and the outer thickness of the DNA all around is about 20 Å, giving a diameter of about 100 Å for the particle overall. The proteins H2A and H2B lie near the ends of the DNA supercoil, while proteins H3 and H4 lie near its center in a compact group. All of these results correspond well with the earlier studies and give more detail: the three-dimensional spool is a sort of wrapped-up version of the string of histones shown in Fig. 7.1(a). Finally, the structure of the histone octamer alone, without DNA, was completed recently at high resolution by B.-C. Wang, E.N. Moudrianakis and colleagues, thereby revealing important details of how the individual histone proteins fit together into a much larger histone octamer.

The pictures shown in Figs 7.1 and 7.2 represent the simplest possible level of organization for DNA and the proteins associated with it in a chromosome. Let us think about the ways by which a string of nucleosomes could fold into some higher-order structure. The regularity of the spacing—200 base-pairs per nucleosome—should be an advantage for the assembly of

H2A   H2B   H3   H4   DNA

Figure 7.2 X-ray diffraction model of DNA wrapped around a histone octamer, shown with top and bottom views. Each number 0, 1, 2, etc. marks one double-helical turn of DNA at low resolution. The DNA is coloured brown, while the histone proteins have more distinctive colors. Courtesy of Tim Richmond.

some sort of ordered fiber, as opposed to an irregularly shaped "clump."

The structure of a polynucleosome fiber is still not known for certain, but one widely accepted model is that of John Finch and Aaron Klug, later extended by Fritz Thoma, Theo Koller and Aaron Klug. In 1979, these scientists reported that one could stabilize the structure of a polynucleosome fiber by soaking it overnight in fixative, prior to taking pictures by electron microscopy. They then saw structures in the microscope that look like the model shown in Fig. 7.3. Unfortunately, the electron microscope does not give the degree of resolution that one would like to have, and so some interpretation is necessary. In their model, individual nucleosomes wrap into a compact spiral that advances by 110 Å per turn. Each turn of the structure contains about 5–6 nucleosomes, depending upon the solution conditions, and the DNA itself adopts a toroidal configuration. Histone H1 is thought to stabilize the folding of nucleosomes into this compact form: in the absence of histone H1, the nucleosomes are not "frozen" by the fixative into some sort of spiral structure, but instead lie like "beads on a string" across the microscope support, as shown at the bottom of the picture. Whether or not histone H1 binds to the DNA depends on the chemical condition of the surrounding fluid, such as the amount of salt present.

The structure shown in Fig. 7.3 is known as a "300 Å fiber," because its outer diameter is typically 250–300 Å. The length of such a fiber is less than that of the constituent string of nucleosomes by a factor of about 6; so the length of the fiber is less than the length of a simple DNA thread by a factor of about 6 × 6, or approximately 40. Many important aspects of this structure, such as the precise location of histone H1, remain undetermined and are a subject of current research.

Finally, we must emphasize that there are other important proteins in the cell nucleus as well as the histones. In rank order after the histones, the next most abundant kinds of DNA-binding protein are the so-called "high-mobility group" or "HMG" proteins. These small proteins were noted by early investigators because they ran quickly through gels in electrophoresis experiments (see Chapter 8), when using an acetic acid–urea buffer. The three main types of HMG protein are known as HMG-1,2, HMG-14,17 and HMG-I,Y. Each class is about 1–10% as abundant as the histones in most tissues. It has long been known that HMG-14,17 are associated with regions of DNA near actively transcribing genes, where the chromatin

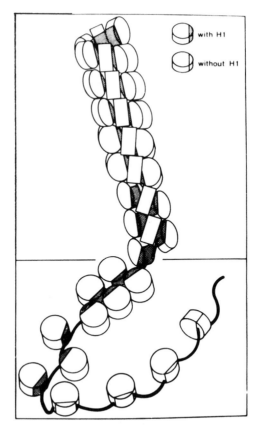

with H1

without H1

Histone octamers can be assembled onto long DNA like "beads-on-a-string," and then these beads can wrap into a flat spiral or "300 Å fiber" in the presence of histone H1. Adapted from F. Thoma *et al.* (1979).

is partly unravelled, but the roles of HMG-1, 2 and HMG-I, Y remain uncertain at present.

We have now summarized some of the more important small-scale features of chromosome structure. Let us therefore proceed to consider the large-scale features. Here the situation becomes very complicated. The problem is that, for most of the time, the DNA and histones are spread so uniformly in the cell nucleus that they show few distinguishing features, and so remain "invisible" to both the light and the electron microscope. In other words, the 300 Å fibers seem to be disposed in the nucleus rather loosely, like a bunch of spaghetti in a bowl, as shown schematically in Fig. 1.2. It is only when a cell is on the point of dividing that these fibers condense or fold into structures that are sufficiently compact to be seen by use of a light or electron microscope. Such highly compact structures are known as "metaphase" chromosomes: some pictures of these were shown in Fig. 1.3. Clearly, the 300 Å fibers within metaphase chromosomes must be packed rather densely around each other into some sort of regular array, although we do not understand at present how this is accomplished.

Yet it seems likely that the 300 Å fibers in their dispersed or "interphase" state are not organized quite so loosely or randomly as spaghetti in a bowl. They must eventually undergo compaction when a cell divides at metaphase, and it is hard to see how they could do this if they were dispersed entirely at random during the intervening periods. Therefore, we should look for an additional level of structure between that of the 300 Å fiber and the condensed, metaphase chromosome.

The strongest evidence for such an intermediate level of organization comes from the strange "polytene" chromosomes which are found in the salivary gland of the fruit fly, and in a few other insects. The fly has four pairs of chromosomes, and by some trick of Nature, each chromosome in the salivary gland can make about 1000 copies of itself during interphase. These copies associate in a side-by-side, parallel fashion to create a highly-ordered structure which can be seen by means of the light microscope. Some very clear, detailed pictures of fly polytene chromosomes were shown in Fig. 1.4. Each individual chromosome retains its dispersed form, but there are so many copies of it in the polytene chromosome, all in register with one another, that the assemblage almost becomes visible to the naked eye, its overall size being about 1/30 mm.

Furthermore, these giant constructions show a great deal of substructure, in the form of well-defined regions of dark and light, or "bands" and "interbands." The bands contain 95% of the total DNA plus protein, while the interbands contain 5%. Thus, DNA and protein must be packed together much more tightly in the bands than in the interbands, because both regions are of about the same size. It would be good to know how the bands and interbands come about. They provide some of the clearest evidence concerning the structural organization of a chromosome between the level of the 300 Å fiber and that of the final, compact form which is seen only when a cell divides.

In Fig. 6.3 we suggested that DNA might sometimes form "loops" of size about 50 000 base-pairs, at somewhat regular intervals along the length of a chromosome. In such a model, each loop would contain an average of about 50 turns of 300 Å fiber, or 250 nucleosomes. (Incidentally, the thin curly line in Fig. 6.3 represents a string of nucleosomes, rather than the DNA itself.) It seems possible that the bands and interbands seen in polytene chromosomes are constructed from a series of such loops. Each loop would compact the DNA longitudinally, thereby providing for the dense packing of DNA and protein seen in the polytene "bands"; while the intervals between loops would correspond to the "interbands," where the packing of DNA and protein is much less dense. Of course, the individual loops themselves cannot be seen in the light microscope, because it takes 1000 nearly identical loops to make one band; but one can see these loops in other unusual sorts of chromosome, as we shall describe below.

Recall that the length of a DNA double-helical thread is reduced by a factor of about 40 when it wraps into a 300 Å fiber. Folding the 300 Å fiber into a series of loops will reduce its length by another factor of about 25, yielding a total compaction of about $40 \times 25 = 1000$, on going from free DNA to the dispersed, interphase form of a typical chromosome. So when a cell divides, it will only have to reduce the size of its chromosomes by a further factor of about 10, in order to reach the total compaction of 10 000-fold mentioned at the beginning of this chapter.

Now are these loops real, or are they just an attractive model, devised to explain phenomena which we cannot understand at present? The most direct evidence in favor of such loops has come from the detailed study of so-called "lampbrush" chromosomes. These lampbrush chromosomes are found in animals such as the frog or newt, in cells that are about to become egg cells. They show very clear "loop" structures of the kind sketched in Fig. 6.3, which can be seen by use of either a light or an electron microscope. The loops in these chromosomes are readily visible for one simple reason: because the genes along the loops are churning out huge amounts of RNA, in preparation for making a new frog egg or newt egg. Thus, these genes are covered almost entirely by RNA polymerase molecules and their associated RNA chains, rather than by histone proteins. The DNA has lost almost all of its compaction due to wrapping about the histone proteins, and yet it remains relatively dense, owing to the great accumulation of protein and RNA along the length of any loop: so it is easy to see such loops in the microscope.

As shown in Fig. 7.4, the polymerase molecules pack very densely along the length of the DNA in these loops from a lampbrush chromosome, like cars in a queue at a traffic light. The RNA chains that emerge to either side are coated in protein (not shown in the diagram), and these chains grow longer as the polymerase molecules travel for greater distances around the loop. Each loop is anchored at its base in two places to certain unknown proteins (or other kinds of molecule) that provide a firm support or "scaffold" for the flexible loop; and then there is an interval of some distance between loops, until another point of attachment to the scaffold is reached. These arrangements are rather similar to the "band–interband" kinds of structure seen in fly polytene chromosomes, only there the loops become much more condensed because they are covered in histone proteins.

Thus, in every case where we can actually see the fine structure of a chromosome by use of the light or electron microscope, we can see evidence for an intermediate level of structure between that of the 300 Å fiber and that of the folded metaphase chromosome. Furthermore, it is not unreasonable to suppose, as a working hypothesis, that this intermediate level of structure might consist simply of a series of large loops in the DNA, together with the intervals between loops. But there are almost no certain data from biochemical studies today to show which sequences in the DNA might attach themselves to the scaffold at the base of every loop, nor to which proteins in such a scaffold the DNA might be attached.

The usual procedure for investigating structures at this level, which has been followed by many workers, has been to remove all the histones from the DNA by use of salt, detergent or other reagents, and then to say that the proteins left represent the

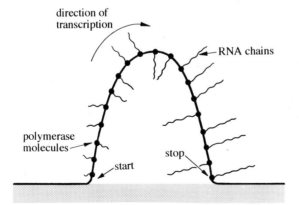

Figure 7.4 RNA polymerase molecules working their way around a loop of DNA in a lampbrush chromosome: schematic for both DNA and the base of the loop, or "scaffold." The RNA chains made by these polymerases grow longer as the polymerases travel further along the loop.

"scaffold." There are obvious dangers associated with this kind of approach, however, and the results are controversial. For example, unless the DNA is attached to the scaffold more firmly than to the histones, the true attachments of DNA to the scaffold may be lost during the treatment. Thus, in doing such an experiment, one might be left at the end with some sort of residual protein, rather than with any scaffold. Although the procedure used by Uli Laemmli provides evidence from the consistency of the DNA sequences protected that the scaffold proteins have remained in their original places, the various procedures used are potentially destructive. Indeed, Dean Jackson and Peter Cook have shown that the apparent size of "loops" obtained by such methods is highly sensitive to the means of preparation of the sample. Furthermore, the loops as mapped by this kind of approach in fly polytene chromosomes do not show any correspondence to the band and interband structures that can be seen clearly by use of a light microscope. Finally, in one case it has been shown that a certain piece of DNA will attach itself to the scaffold only after, and not before, the addition of detergent or similar reagent.

Some very interesting work on this difficult subject has been done in Siberia, by V.F. Semeshin and colleagues. They have studied by electron microscopy the insertion of foreign bits of DNA into a fly polytene chromosome. For their foreign DNA, they used a special piece of DNA known as a "P-element," that contains sequences which enable a fly enzyme to splice foreign DNA into fly DNA. They find that the P-element can either make a new band, or else split a previously existing band, depending upon its site of insertion in the chromosome. There are also other cases, not yet studied by electron microscopy, where the gene activity within a P-element can be "insulated" from its position in a chromosome by attaching certain DNA sequences to both sides. Normally the activity of any gene is highly sensitive to its position in a chromosome, yet these experiments suggest that certain sequences in the DNA can set up a "boundary," possibly in the form of a loop-attachment site, between different genes on a chromosome.

By this point you may be feeling frustrated at the general air of uncertainty in our presentation. Ever since we stopped talking about nucleosomes, and went on to talk about the 300 Å fiber, HMG proteins, loops and scaffolds, there has hardly been anything definite to learn. But before we leave this topic, we shall mention one further aspect of the looping behavior of DNA in chromosomes, which seems to have a lot to do with how genes work; it also illustrates only too well how fluid is our knowledge of these important matters. In Fig. 7.4 we implied that the loops of DNA in a chromosome might be rigidly-fixed structures, because the DNA in each loop seems to be held at its two ends by a protein rod or scaffold. But only some of the scientists working on chromosomes today think in that way. The others think that the loops are flexible structures, that allow the DNA to slide or thread itself through the base of a chromosome, in the same way that a piece of magnetic tape goes through the reading-head of a tape recorder.

The two alternative models are sketched in Fig. 7.5. In Fig. 7.5(a), the loops are drawn as if they are fixed objects, and it is assumed that RNA polymerase travels along the genes which are contained in the outer parts of each loop. In other words, the polymerase and its associated proteins start making RNA near one end of the loop, and stop when (or before) they reach the other end. In Fig. 7.5(b), the loops are able to slide through the base of the chromosome, where it is supposed that all of the polymerases and their associated proteins are stored in a kind of "active compartment." The DNA then threads itself through this active region so as to come into contact with the polymerase, and thus to make RNA.

At present, there is some evidence in support of each of these two theories. You must realize that biologists today are able to isolate many of the proteins that make genes work (the so-called "transcription factors"), but they do not yet know where these proteins come from—whether from the base or the tip of a chromosomal loop. On the one hand, it is known that a polymerase molecule can track around the outer parts of a loop in special cases (see Fig. 7.4), when the genes are making very large quantities of RNA. On the other hand, some loops within a lampbrush chromosome are thought to be able to change size, as shown schematically in Fig. 7.5(b). Also, in relation to normal chromosomes that are not making large quantities of RNA, it has been reported that one can cut away 90% of the DNA and protein from the outer parts, to leave just the central part or "scaffold;" and in the process one keeps almost all of the enzymatic activities needed for making RNA.

We personally do not favor one or the other of these different models for chromosome structure. We have described them here in order to show you that the issue is not settled, and that nobody yet knows for certain how the genes work in our chromosomes.

After this excursion into looping and gene activity, let us now

Figure 7.5 Two hypotheses for the location of gene activity in a typical chromosome from higher organisms, and here shown schematically. We do not know if the hypothetical loops of DNA are rigidly fixed structures as in (a), or if they are flexible enough to let the DNA slide through their sites of attachment to a "scaffold" as in (b).

return to better-understood subjects. On a very much larger scale than we have considered so far, each chromosome has two kinds of specialized structure, which correspond to specific sequences in the DNA. These are known as the "centromere" (or "central part") and the "telomeres" (or "end parts"), as shown schematically in Fig. 7.6. The centromere lies somewhere within the main part of each chromosome, and it is the feature which becomes attached to the tubular protein structures known as "microtubules," which assemble themselves when the cell is about to divide. These microtubules pull the duplicated chromosomes apart, thus providing one copy of each chromosome for each new cell. In the photograph of Fig. 1.3, the two DNA centromeres lie within the narrow, central X-shaped part of each duplicated chromosome. The telomeres, on the other hand, lie at either end of a long, linear chromosome, and their role remains uncertain. Obviously they can "seal" the ends of any chromosome to prevent its joining to other chromosomes; but they may well do more than that. Some scientists think that the telomeres adhere to the nuclear membrane, so as to anchor the chromosomes in three-dimensional space. Other people think that the telomeres are needed to assist in the copying of a linear chromosome upon cell division, because the usual copying enzyme (DNA polymerase) cannot copy the DNA all the way to the end of a linear molecule, on both strands of the double helix: a special enzyme, known as "telomerase," copies the telomeres.

Let us now consider in more detail the structure of a compact, metaphase chromosome. Many studies by light and electron microscopy have shown that such duplicated chromosomes, which are ready to be separated into two parts on cell division (where one copy goes to each daughter-cell), have a sort of spiral structure as shown in Fig. 7.7(a). You can see there the same overall shape as in Fig. 1.3, but now more detail is visible. A thick rod or sausage-shape coils round and round to make spiral arms that serve to reduce the overall size of the chromosome to assist in cell division. Of what, precisely, might this thick rod be made? One obvious possibility would be the scaffold-and-loop structure which we have already described (Fig. 7.7(b)). The loops are small in this picture by comparison with those shown earlier, because the two diagrams in Fig. 7.7(b), which provide both side and end-views of the loops, are drawn on the scale of the much larger Fig. 7.7(a). Yet it seems possible that the bulk of the spiral rod may consist mainly of these loops, while the protein scaffold constitutes perhaps a relatively narrow core.

The formation of compact coils from a scaffold-and-loop structure would account nicely for the final compaction by a factor of 10, that is needed to achieve an overall compaction of 10 000 between the length of the free DNA and the length of a metaphase chromosome. It seems possible that the cell makes

Figure 7.6 Special functional regions of chromosomal DNA. A centromere need not necessarily be located at the very center of the chromosomal DNA, but the telomeres always occupy the ends. Telomeres are short, repetitive DNA sequences such as TTAGGG, repeated hundreds of times in a row (in fish, frogs, and humans). Centromeres are long, semi-repetitive sequences of repeat-length at least 1000–10 000 base-pairs, that may provide for multiple attachment sites of DNA to microtubules. Many of the chromosomal centromeric regions have been characterized and isolated from simple organisms such as yeast, but no functional centromere has yet been isolated from any higher organism.

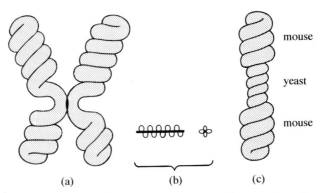

Figure 7.7 (a) Detailed fine-structure of the duplicated, metaphase chromosome as a wrapped-up spiral-rod or sausage-shape (compare with Fig. 1.3). (b) This same rod shown in detail as a hypothetical scaffold with loops, in two views. (c) A special mouse–yeast hybrid chromosome of two different diameters, revealing our ignorance of the factors that influence chromosome structure on a large scale.

some special protein that induces curvature in the scaffold-and-loop structure, as the cell gets ready to divide, thereby providing for the change from an interphase to a metaphase chromosome. Both left-handed and right-handed spirals of the kind shown in Fig. 7.7(a) have been seen by light and electron microscopy.

Scientists have recently found many different ways of transferring the DNA from one organism into the chromosomes of another. For example, the total DNA from a yeast chromosome of 9 million base-pairs can be inserted into the chromosome of a cell isolated from the mouse by "fusion" of the two kinds of cell in tissue culture. The resulting metaphase chromosome has a strange, dumb-bell shape (Fig. 7.7(c)): this picture is a simplified version, corresponding to one half of Fig. 7.7(a). The key feature is that the central part, which contains the yeast DNA, has only half the diameter of a normal mouse chromosome. One possible explanation is that the loops of yeast DNA are shorter than those of mouse DNA, thereby making the scaffold-and-loop structure smaller in diameter. Another possibility is that the protein scaffold in yeast curves into tighter spirals than does the scaffold in mouse. And perhaps both of these effects happen simultaneously. In any case this is a very deep result, because it reveals that there is a definite level of organization in a metaphase chromosome, as specified by the sequence of DNA, which we do not currently understand.

Eventually, and probably in the next century, people will want to assemble full-length, authentic chromosomes in a test-tube, and to use these in agriculture (and perhaps medicine) in order to do useful things that are not yet possible. Most present-day agricultural plants and farm animals are not wild species: instead, they are species that have been selectively bred for their food value by farmers over thousands of years. For example, the well-known "Granny Smith" apple and the "Parson Brown" orange were bred by people in the last century for improved quality of fruit. Very recently, some scientists at the Calgene company in California have made a tomato plant which produces tomatoes that do not rot on long storage, but remain firm and red. This was done by adding to the tomato plant small amounts of foreign DNA which inhibit production of the "rotting" enzyme. This kind of research will no doubt continue, and it will require the insertion of larger and larger bits of DNA into various plants and animals. Eventually it may be necessary to add such DNA in the form of folded chromosomes, because the naked, protein-free DNA is long and thin and easily broken, and much longer than the size of the cell itself.

So far, scientists have made only the smallest start at understanding the assembly of DNA into chromosomes. They would like to be able to put into a test-tube moderate quantities of pure DNA, pure protein, and perhaps other substances, in order to assemble chromosomes in the laboratory. But this sort of thing is still a long way off. At present, histones and DNA can only be combined to make authentic nucleosomes in the test-tube by adding "extracts" taken from living cells. These extracts are presumed to contain many important factors for the assembly of nucleosomes, and perhaps for the assembly of loops and scaffolds; but it is not yet known exactly what these factors are, or how they work. Some of the extracts can even assemble a nuclear membrane around the DNA. There is a lot of interesting work to be done here.

The most important early result in this general area was obtained by Ron Laskey in 1977. He found that one could incubate DNA with an extract from frog eggs, in order to make authentic fragments of a chromosome. Eventually he and Juergen Kleinschmidt independently found two proteins in the frog egg, called "N1" and "nucleoplasmin," that bind to the histone pairs H3,H4 and H2A,H2B, respectively, and carry them onto the DNA. Why should "carriers" such as these be needed to place the histones onto the DNA in the form of nucleosomes? Why should histones not be able to bind to DNA spontaneously? The water in our cells contains various dissolved salts, at low concentration. In the test-tube, histones place themselves on DNA in the form of nucleosomes only if the salt concentration is first made much higher than it is in our cells, and is then reduced slowly. At physiological, cell-like salt concentrations, the positively charged histones tend to aggregate in clumps, rather than form nucleosomes. But the carrier proteins N1 and nucleoplasmin contain many negatively charged amino acids, such as aspartate and glutamate; and these bind tightly to the histones so as to prevent them from aggregating. In fact, a simple polymer of aspartate or glutamate can also assemble histones onto DNA. Nature uses complicated proteins such as N1 and nucleoplasmin instead for two reasons: first, to carry the histones from their site of synthesis on the ribosome, across the nuclear membrane to their site of assembly on the chromosome; and second, to pick out H3,H4 and H2A,H2B from all the different positively charged proteins in the cell that might be able to bind to DNA non-specifically.

The discovery of these histone carriers was good progress, but then a stumbling block was reached. When the complexes of N1 with H3,H4, and nucleoplasmin with H2A,H2B were purified from the cell extract, it turned out that the histones would assemble onto DNA in the test-tube at a spacing of 145 base-pairs, rather than at a spacing of 180 base-pairs as in the original extract from frog eggs. In other living cells, the spacing can vary from as low as 170 to as high as 260 base-pairs, depending on the proteins present in any nucleosome; but one never sees a very short spacing of 145 base-pairs in Nature.

Figure 7.8 shows some typical results for the assembly of histones onto DNA by an extract from living cells on the one hand, and by a pure system on the other. The procedure for determining the spacing of nucleosomes is to "digest" the preparation by means of a DNA-cutting enzyme (like the one used by Hewish and Burgoyne), and then to measure the size of the fragments so obtained by using an electrophoretic gel according to a scheme which we shall describe in Chapter 8.

On the left-hand side of Fig. 7.8, one can see by "reading" the gel that DNA may be assembled with histones from a frog-egg extract to make particles of size 180, 360, 540, 720 and 900 base-pairs; and so the size of the fundamental unit is 180 base-pairs. On the right-hand side of Fig. 7.8, DNA has been assembled with histones in a pure system (that is, without any cell extract), and it has made particles of size 145, 290, 435, 580, 725, and 870 base-pairs, after digestion with the enzyme; and so the size of the fundamental unit is 145 base-pairs. (The lowest band of size 145 base-pairs is obscured in this gel, but it can be seen in other experiments.)

Figure 7.8 Analysis of nucleosome assembly by use of electrophoresis in gels, after digestion by enzymes. Histones are assembled onto DNA at a spacing of 145 base-pairs in a pure system, but at a more authentic spacing of 180 base-pairs in an extract from frog eggs. Left-hand gel by courtesy of David Tremethick.

One possible explanation for the reduction in spacing from 180 to 145 base-pairs, on going from a cell extract to a pure system, is that we have lost certain important proteins when purifying the material. If these proteins normally associate with the histones as they assemble onto DNA in a cell, then in the cell extract there would be more proteins per particle than in the pure system, and hence each particle would extend for a greater length along the DNA. Which proteins might we have lost? The frog-egg extract is known not to contain any histone

H1, so perhaps some other abundant protein (or proteins) might be important to the correct assembly of nucleosomes.

The most likely candidate proteins for this task would be the high-mobility group proteins HMG-1,2, HMG-14,17 and HMG-I,Y, which we mentioned earlier.[1] These small proteins are fairly abundant in the cell nucleus, and they also bind to DNA with low specificity for the base sequence. Recent experiments by David Tremethick have shown that a moderately pure fraction of proteins taken from the frog-egg extract will increase the spacing of nucleosomes in a pure system from 145 to 170 base-pairs, even though this fraction does not contain histone H1, or indeed any of the four histones H2A, H2B, H3, and H4. Clearly, this would seem to be evidence that some spacing protein "X," other than histone H1, may be involved in the proper assembly of nucleosomes. Spacings of 170 base-pairs are also seen in yeast cells, which contain the four histones H2A, H2B, H3, and H4, but not histone H1, so perhaps protein X is present in yeast as well.

When histone H1 is added to the frog-egg extract, or to a partly purified extract, the spacing of nucleosomes grows even larger, from about 170 or 180 to about 200 or 220 base-pairs, an increase of 30–40 base-pairs. Independent experiments by Antonio Rodríguez-Campos and Tremethick show this clearly. Most mature tissues in animals such as the fly, chicken and human contain histone H1, and they also show spacings on the gel which correspond to a fundamental unit from 200 to 220 base-pairs long. Animals such as the sea urchin show a spacing of 260 base-pairs in some tissues, and perhaps these cells contain other special nucleosome-assembly proteins that have not yet been identified.

A tentative summary of these results is shown schematically in Fig. 7.9. It seems likely that the assembly of DNA into nucleosomes proceeds by at least three steps. First, the N1 protein (or some other negatively charged carrier) binds two copies each of histones H3,H4 and assembles them onto DNA. Second, the nucleoplasmin protein may bind H2A,H2B, and possibly protein X, and add these proteins to DNA on either side of the already-assembled H3,H4 tetramer. Without protein X there may only be eight proteins per particle, spanning 145 base-pairs of DNA; but with protein X there are perhaps ten proteins per particle, spanning 170 base-pairs. Finally, if histone H1 is present, it may add to the DNA between the particles, to increase the spacing from 170 to about 200 base-pairs, as in a typical mature tissue from an animal or plant. The carrier for histone H1 in living cells is not known, but in the extract it seems to be some kind of RNA. No doubt the conjectured picture shown in Fig. 7.9 will be altered and refined in the future, but at this stage it seems useful to have a picture of this sort which fits some of the data, rather than no picture at all.

One important consequence of a stepwise assembly of DNA into nucleosomes, as in Fig. 7.9, is the following: each step in the procedure is reversible, so we can expect disassembly to proceed by the same series of steps but in a reverse direction. The relative rates of assembly versus disassembly within any step will therefore be of importance to specific biological processes, such as gene activation, that require the DNA to be at least partly unravelled from its folded state. We shall return to this subject later.

A large part of this book has been devoted to a study of DNA curvature. How does the curvature of DNA relate to its assembly into nucleosomes? In Figs 7.1 and 7.2, it was shown that DNA curves for almost two superhelical turns, each of about 80 base-pairs, around the histone proteins in any nucleosome. Thus, one might expect that curved or else very flexible DNA could assemble into nucleosomes more easily than DNA of mixed sequence, because it would take less energy for these special

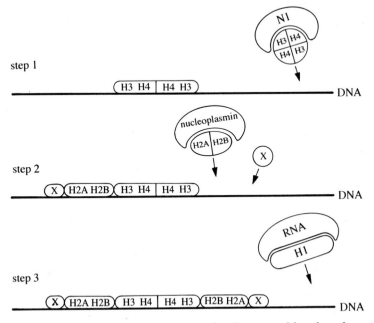

Figure 7.9 Three-step hypothetical scheme for the assembly of nucleosomes. The identity of the protein X is not yet known; it is a partially purified protein fraction from a frog-egg extract.

sequences to wrap into the required superhelical shape. In many experiments, it has been found that DNA which is curved on account of its repeating sequence does indeed bind more tightly to the histone proteins than other, non-curved DNA. Generally, even a small amount of curvature (or anisotropic flexibility), as specified by the DNA base sequence, is sufficient to locate the path of the DNA about the histone proteins. In addition, experiments using living cells have shown that the histone proteins often adopt highly ordered locations with respect to the DNA in a cell nucleus in a way that is influenced, but not strictly determined, by DNA flexibility. One can calculate the ease of curvature of any DNA sequence about a single set of histone proteins by algorithms similar to those described in Chapters 4 and 5. Yet it is difficult to calculate such things precisely over several successive sets of histone proteins, because successive nucleosomes may stick together, or fold into a 300 Å fiber, in a cooperative fashion that works against preferences in the DNA for curvature or flexibility.

Several experiments have shown that the insertion of curved or else very flexible DNA near the start of a gene can actually repress the activity of that gene. In these cases it seems likely that the histone proteins bind to the DNA more tightly than usual, on account of curvature or flexibility in the DNA; and so the specialized "transcription factor" proteins which are used to initiate RNA synthesis cannot gain access to the DNA, in order to make the gene work. (The zinc-finger Zif268 protein described in Chapter 4 is one example of such a transcription factor.) Furthermore, when curved DNA is present on large scale throughout an entire chromosome (as, for example, in certain birds and reptiles), then genes throughout the whole chromosome may be repressed, and the chromosome itself may remain so compactly folded and condensed that it becomes visible at interphase.

A broad conclusion which can be drawn from these studies is that the actions of DNA in a chromosome are very intricate and complex. If you try to study the workings of DNA solely in terms of abstractly defined genes and transcription factors, without paying attention to the structure of the DNA, and to the proteins which are tightly associated with the DNA in a chromosome, then you will encounter many cellular phenomena which you will be unable to explain. The vast majority of cell-free studies done today on the mechanisms of transcription in higher organisms use naked DNA rather than DNA packaged into authentic pieces of chromosome. One can only wonder whether the results obtained from such studies are truly relevant to the mechanisms of transcription in living cells.

A major precursor to the activation of any gene must be, of course, the unravelling of DNA from its tight folding about the histone proteins. Without such an unravelling, the RNA polymerase and other proteins required for "reading" the DNA could never gain access to their required binding sites near the start of a gene. How does a cell control the assembly or disassembly of DNA into folded structures in the vicinity of specific genes? No one knows for certain, but one likely means is by "protein modification." The histones, the HMG and many other proteins can be "chemically modified" in several different ways, the most important of which are "acetylation" and "phosphorylation." These modifications are performed by special enzymes known as "acetylates" and "kinases," respectively. When a protein is acetylated, then an acetyl group (CH_3CO) is added to one or more of its lysine amino acid residues, thereby removing the positive charge (Fig. 7.10(a)). When a protein is phosphorylated, then a phosphate group (PO_3^-) is added, usually to one or more of its serine or threonine residues, thereby adding negative charge (Fig. 7.10(b)). We can see that these protein modifications are a way of altering the electric charge of a protein; and indeed, it is well-known that such modifications may be made by special enzymes even after a protein has been built into the structure of a chromosome.

The most important kinds of protein modification, in the context of chromosome assembly, are listed in Table 7.1. Histones H3,H4 and HMG-1,2 can be acetylated. The acetylation of H3, H4 makes these proteins more soluble when they are bound to

Figure 7.10 (a) Acetylation and (b) phosphorylation of typical amino acids in a protein. Both of these modifications change the electric charge of the protein, in a way that may influence its structure and its ability to interact with DNA or with other proteins.

DNA, on account of the reduced positive charge; and it is known from experiment that DNA which is bound to acetylated H3 and H4 remains partly active towards RNA synthesis—that is, it can be transcribed, whereas the full complement of histones greatly inhibits transcription. So we may not have to remove all of the histone proteins from DNA in order to permit gene activity; perhaps we can just remove H1, H2A, and H2B, and also acetylate H3,H4, and the gene will be active.

Many different chromosomal proteins can be phosphorylated, including HMG-14,17, HMG-I,Y, histone H1, the carriers N1 and nucleoplasmin, and other minor proteins such as P1 and D1. The phosphorylation of HMG-14 causes it to bind to H2A and H2B more tightly than otherwise; but it is not yet known if this helps HMG-14 to remove H2A and H2B from DNA. The consequences of HMG-17 phosphorylation are similarly not yet known. HMG-14,17 are found near active genes, and one wonders whether these proteins must be phosphorylated to induce gene activity by unravelling the chromosome. Also, it is not known what biological effects might result from the phosphorylation of HMG-I,Y or of P1 or D1, even though these proteins are perhaps the major targets for kinases in the cell nucleus. The phosphorylation of histone H1 causes it to bind DNA less tightly; so in this case phosphorylation will definitely unravel the chromosome locally, if it helps H1 to be removed. (Remember that H1 stabilizes the 300 Å fiber.) Finally, the carriers N1 and nucleoplasmin need to be phosphorylated to

Table 7.1 The acetylation or phosphorylation of certain chromosomal proteins can affect their physical properties, and so influence their biological functions

Protein	Modification	Effect
H3,H4	Acetylation	Easy sliding on DNA, greater solubility
HMG-1,2	Acetylation	Forms tetramer
HMG-14,17	Phosphorylation	Binds H2A, H2B more tightly
HMG-I,Y	Phosphorylation	Not known
P1 or D1	Phosphorylation	Not known
H1	Phosphorylation	Binds DNA less tightly
N1, nucleoplasmin	Phosphorylation	Bind histones more tightly

have enough negative charge to bind the histones tightly, and so to be active in assembly of histones onto DNA.

Thus, the unravelling of a chromosome near a specific gene can perhaps be assisted through the phosphorylation of histone H1, the phosphorylation of HMG-14 or other proteins, and the acetylation of H3 and H4. The changes in electric charge on these proteins may modify the relative rates of assembly or disassembly of the three steps shown tentatively in Fig. 7.9. Unfortunately, there are many different enzymes in the cell nucleus which carry out acetylation or phosphorylation of the chromosomal proteins, and it is hard to sort out which enzyme modifies which protein and to what end. One interesting feature is that these chromosome-modification enzymes, both the acetylases and the kinases, do not float freely within the nucleus, but are generally bound somewhere to DNA. It seems likely that these enzymes will be found near active genes, where they might help to unravel the chromosome; but no experiments on this subject have yet been done with any firm conclusions. Nevertheless, it is known that the H3 and H4 proteins in nucleosomes are highly acetylated when present in active genes, and also that nucleosomes are acetylated in clusters, as if all of the nucleosomes near an acetylase enzyme have become chemically modified.

Now we have covered almost all major aspects of current research into chromosome structure and gene activity. It is sad to say that we do not yet know very much. What of the future? One hopes that the next 20 years will bring a fresh generation of scientists to tackle some of the problems mentioned above. Perhaps some of you reading this book may decide to study chromosome structure, now that you are aware of our current state of ignorance, and of the great significance of the results that might be obtained. In the very distant future, we can perhaps wonder whether any of the great problems in biology, such as the growth and development of higher organisms, will be solved by a deeper understanding of chromosomes.

Note

1. Recent work by one of us (H.R.D.) shows that HMG-14,17 can act as a spacing protein "X" in a pure assembly system consisting just of histone octamers, HMG protein, and DNA.

Further Reading

Allshire, R.C., Cranston, G., Gosden, J.R., Maule, J.C., Hastie, N.D., and Fantes, A.P. (1987) A fission yeast chromosome can replicate autonomously in mouse cells. *Cell* **50**, 391–403. The yeast–mouse hybrid chromosome having two different diameters (shown schematically in Fig. 7.7) came from cell-line F1.1 of this paper.

Almouzni, G., Mechali, M., and Wolffe, A.P. (1991) Transcription complex disruption caused by a transition in chromatin structure. *Molecular and Cellular Biology* **11**, 655–65. Histones H3 and H4 alone on DNA allow for high rates of transcription.

Blackburn, E.H. (1991) Telomeres. *Trends in Biochemical Sciences* **16**, 378–81. An up-to-date summary of telomere research.

Callan, H.G. (1982) Lampbrush chromosomes. *Proceedings of the Royal Society of London,* **B 214**, 417–48. Many beautiful pictures of scaffolds and loops.

Cook, P.R. (1989) The nucleoskeleton and the topology of transcription. *European Journal of Biochemistry* **185**, 487–501. Are genes located on the outer parts of loops, or at their bases?

Costanzo, G., DiMauro, E., Salina, G., and Negri, R. (1990) Attraction, phasing and neighbor effects of histone octamers on curved DNA. *Journal of Molecular Biology* **216**, 363–74. Histone proteins bind more tightly to curved than to non-curved DNA.

Eggert, H. and Jack, R.S. (1991) An ectopic copy of the *Drosophila ftz*-associated SAR neither reorganizes local chromatin structure nor hinders elution of a chromatin fragment from isolated nuclei. *EMBO Journal* **10**, 1237–43. The DNA that contains a supposed "loop-attachment site" becomes part of the "scaffold" only after and not before the addition of detergent.

Espel, E., Bernués, J., Guasch, M.D., Querol, E., Plana, M., and Itarte, E. (1987) Phosphorylation of high-mobility-group protein 14 by two specific kinases modifies its interaction with histone oligomers in free solution. *Biochimica et Biophysica Acta* **909**, 190–200. Phosphorylation of HMG-14 lets it bind tightly to histones H2A,H2B.

Hill, C.S., Rimmer, J.M., Green, B.N., Finch, J.T., and Thomas, J.O. (1991) Histone–DNA' interactions and their modulation by phosphorylation of Ser-Pro-X–Lys/Arg motifs. *EMBO Journal* **10**, 1939–48. Studies of how the phosphorylation of H1 reduces its binding to DNA.

Hill, R.J., Mott, M.R., and Steffensen, D. (1987) The preparation of polytene chromosomes for localisation of nucleic acid sequences, proteins, and chromatin conformation. *International Review of Cytology* **108**, 61–118. A review of work on polytene chromosomes.

Kellum R., and Schedl, P. (1991) A position-effect assay for boundaries of higher order chromosomal domains. *Cell* **64**, 941–50. Certain DNA sequences can insulate the transcriptional activity of one gene from that of its neighbors along the length of a chromosome.

Lund, T., Skalhegg, B.S., Holtlund, J., Blomhoff, H.K., and Laland, S.G. (1987) Fractionation and identification of metaphase-specific phosphorylated forms of high-mobility-group proteins. *European Journal of Biochemistry* **166**, 21–6. Techniques for isolation and purification of the HMG proteins in a variety of phosphorylated forms.

Mirkovitch, J., Spierer, P., and Laemmli, U.K. (1986) Genes and loops in 320,000 base-pairs of the *Drosophila melanogaster* chromosome. *Journal of Molecular Biology* **190**, 255–8. No correlation is found between loops as measured by histone depletion, and bands as seen by light microscopy in polytene chromosomes.

Rattner, J.B. and Lin, C.C. (1985) Radial loops and helical coils coexist in metaphase chromosomes. *Cell* **42**, 291–6. Pictures of duplicated metaphase chromosomes as made from a spiral rod.

Saitoh, Y., Saitoh, H., Ohtomo, K. and Mizuno, S. (1991) Occupancy of the majority of DNA in the chicken W chromosome by bent-repetitive sequences. *Chromosoma* **101**, 32–40. Curved DNA in a chicken chromosome.

Schulman, I. and Bloom, K.S. (1991) Centromeres: an integrated protein/DNA complex required from chromosome movement. *Annual Reviews of Cell Biology* **7**, 311–36. A good review of centromeres in yeast and human beings.

Sessa, G. and Ruberti, I. (1990) Assembly of correctly spaced chromatin in a nuclear extract from *Xenopus laevis* oocytes. *Nucleic Acids Research* **18**, 5449–55. First evidence for a spacing protein "X" that needs magnesium and ATP to work.

Straka, C. and Horz, W. (1991) A functional role for nucleosomes in the repression of a yeast promoter. *EMBO Journal* **10**, 361–8. DNA sequences that bind tightly to the histone proteins can repress gene activity, by blocking the binding of transcription factors.

Turner, B.M. (1991) Histone acetylation and control of gene expression. *Journal of Cell Science* **99**, 13–20. A review of histone acetylation and its effects on gene activity.

Westermann, R. and Grossbach, U. (1984) Localisation of nuclear proteins related to high mobility group protein 14 (HMG 14) in polytene chromosomes. *Chromosoma* **90**, 355–65. The HMG-14 protein is found mainly where genes are making large amounts of RNA.

Wolffe, A.P. (1991) Developmental regulation of chromatin structure and function. *Trends in Cell Biology* **1**, 61–6. A survey of how histone proteins may be chemically modified.

Zinkowski, R.P., Meyne, J., and Brinkley, B.R. (1991) The centromere–kinetochore complex: a repeat subunit model. *Journal of Cell Biology* **113**, 1091–110. A centromere from a mammalian chromosome is shown to consist of many structural repeats, each of which attaches separately to microtubules on cell division.

Bibliography

Arents, G., Burlingame, R.W., Wang, B.-C., Love, W.E., and Moudri-anakis, E.N. (1991) The nucleosomal core histone octamer at 3.1 Å resolution: a tripartite protein assembly and a left-handed superhelix. *Proceedings of the National Academy of Sciences, USA* **88**, 10148–52. High resolution pictures of the histone proteins without the DNA.

Dilworth, S.M., Black, S.J., and Laskey, R.A. (1987) Two complexes that contain histones are required for nucleosome assembly *in vitro*: role of nucleoplasmin and N1 in *Xenopus* egg extracts. *Cell* **51**, 1009–18. Identification of the histone carriers N1 and nucleoplasmin.

Finch, J.T., Lutter, L.C., Rhodes, D., Brown, R.S., Rushton, B., Levitt, M., and Klug, A. (1977) Structure of nucleosome core particles of chromatin. *Nature* **269**, 29–36. First crystals of the nucleosome core particle, comprising the histone octamer plus 145 base-pairs of DNA, and their structural analysis at low resolution.

Gasser, S.M. and Laemmli, U.K. (1987) A glimpse at chromosomal order. *Trends in Genetics* **3**, 16–22. A summary of the many methods used to remove the histones from DNA, in order to study the "scaffold," and discussion of the results obtained.

Hewish, D. and Burgoyne, L. (1973) Chromatin sub-structure: the digestion of chromatin DNA at regularly spaced sites by a nuclear deoxyribonuclease. *Biochemical and Biophysical Research Communications* **52**, 504–10. Discovery of a regular spacing of histone proteins along the DNA in chromatin.

Jackson, D.A., Dickinson, P., and Cook, P.R. (1990) The size of chromatin loops in HeLa cells. *EMBO Journal* **9**, 567–71. The "loop sizes" obtained on histone depletion depend strongly on the exact procedures used.

Kleinschmidt, J.A., Fortkamp, E., Krohne, G., Zentgraf, H., and Franke, W.W. (1985) Co-existence of two different types of soluble histone complexes in nuclei of *Xenopus laevis* oocytes. *Journal of Biological Chemistry* **260**, 1166–76. Independent discovery of N1 and nucleoplasmin by these workers.

Kornberg, R.D. and Thomas, J.O. (1974) Chromatin structure: oligomers of the histones; Kornberg, R.D. (1974) Chromatin structure: a repeating unit of histones and DNA. *Science* **184**, 865–8; 868–71. Proposal for a histone octamer as the fundamental unit of protein–DNA assembly in chromosomes.

Laskey, R.A., Mills, A.D., and Morris, N.R. (1977). Assembly of SV40 chromatin in a cell-free system from *Xenopus* eggs. *Cell* **10**, 237–43. Discovery of a cell extract that will make authentic pieces of a chromosome.

Lutter, L.C. (1978) Kinetic analysis of deoxyribonuclease I cleavages in the nucleosome core: evidence for a DNA superhelix. *Journal of Molecular Biology* **124**, 391–420. A good enzyme-digestion study which provided solution evidence for a model in which $1\frac{3}{4}$ turns of superhelical DNA wind in a left-handed sense around the histone octamer.

Olins, A.L. and Olins, D.E. (1974) Spheroid chromatin units (*v* bodies). *Science* **183**, 330–1. First pictures of beads-on-a-string for histones and DNA in chromosomes.

Richmond, T.J., Finch, J.T., Rushton, B., Rhodes, D., and Klug, A. (1984) Structure of the nucleosome core particle at 7 Å resolution. *Nature* **311**, 532–7. Medium-resolution picture of the DNA wrapped about the protein as in Fig. 7.2.

Rodríguez-Campos, A., Shimamura, A., and Worcel, A. (1989) Assembly and properties of chromatin containing histone H1. *Journal of Molecular Biology* **209**, 135–50. Histone H1 increases the spacing in a frog-egg extract from 180 to 220 base-pairs per nucleosome.

Semeshin, V.F., Demakov, S.A., Perez-Alonso, M., and Zhimulev, I.F. (1990) Formation of interdiscs from DNA material of the P-element in *Drosophila* polytene chromosomes. *Genetika (USSR)* **26**, 448–56. The P-element DNA can make new bands or else split old ones.

Thoma, F., Koller, T., and Klug, A. (1979) Involvement of histone H1 in the organization of the nucleosome and of the salt-dependent super-structures of chromatin. *Journal of Cell Biology* **83**, 403–27. Electron microscope study of long arrays of nucleosomes in a chemically-fixed condition. The source of Fig. 7.3.

Tremethick, D. and Frommer, M. (1992) Partial purification from *Xenopus laevis* oocytes of an ATP-dependent activity required for nucleosome spacing *in vitro*. *Journal of Biological Chemistry* (in press). How to isolate the spacing protein X in partly pure form.

Williamson, R. (1970) Properties of rapidly labeled deoxyribonucleic acid fragments isolated from the cytoplasm of primary cultures of embryonic mouse liver cells. *Journal of Molecular Biology* **51**, 157–68. The 200-base-pair spaced pattern of chromatin, but not interpreted as such.

Exercises

7.1 A piece of double-helical DNA, 3010 base-pairs long, is bound to a series of histone octamers in the manner of Figs 7.1, 7.2 and 7.3. Then an enzyme like the one used by Hewish and Burgoyne (1973) is added to a sample. Fine-mapping experiments show that the enzyme cuts only at certain locations along the length of the DNA molecule, which are known to be the empty spaces or "linkers" between histone octamers. These locations, measured in terms of the distance from one end of the 3010-bp DNA, were found by experiment to be (in units of base-pairs): 220, 430, 670, 870, 1090, 1320, 1520, 1740, 1930, 2140, 2350, 2580, 2790.

a Make a table of all possible DNA fragment lengths, for fragments containing either 1, 2, 3, or 4 histone octamers. This table should have four columns. Column 1 should list all fragment lengths of size 220 − 0 = 220, 430 − 220 = 210, etc. base-pairs, and should contain 14 numbers. Column 2 should list fragment lengths of size 430 − 0 = 430, 670 − 220 = 450, etc., and should contain 13 numbers. Column 3 should list fragments such as 670 − 0 = 670, 870 − 220 = 650, and contain 12 numbers. Finally, column 4 should list sizes of 870 − 0 = 870, 1090 − 220 = 870, etc., and contain 11 numbers.

What is the mean fragment size in each of columns 1, 2, 3, 4, and its standard deviation?

b What is the mean spacing of histone octamers in this particular sample? Divide each mean fragment size by the column number 1, 2, 3, or 4 to get an optimal value.

7.2 The model of the histone octamer and its associated DNA, that is shown in Fig. 7.2, may be represented crudely as a protein cylinder of diameter 60 Å and height 50 Å, around which are wrapped two complete turns of DNA having a diameter of 20 Å.

a Calculate the proportion of the total volume which is occupied by the DNA. For this rough calculation you may treat the DNA as two separate hoops, each with an inner diameter of 60 Å. Recall that the volume of a hoop is equal to the product of its cross-sectional area and its mid-circumference.

b Calculate the number of base-pairs in each hoop, assuming that the DNA has a length of 3.3 Å per base-pair.

7.3 The packing of nucleosomes into the "300 Å fiber," as shown in Fig. 7.3, may be represented approximately by diagram (a), below. Here, to make the calculations especially simple, we have drawn the nucleosomes as closed rings of six, rather than as helical spirals with six nucleosomes per turn.

By considering each nucleosome as a cylinder of radius 50 Å and length 50 Å (for the protein plus DNA: see Exercise 7.2), estimate a the outer diameter of each ring, b the diameter of the inner hole, and c the height of the 300 Å fiber per ring of six nucleosomes.

A second possible packing scheme, also expressed in terms of closed rings, is shown in diagram (b) below. The nucleosomes are cylinders, just as before, but now their flat faces lie

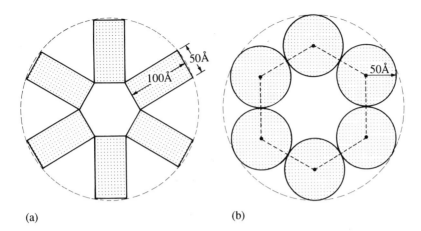

(a)　　　　　　　　　　　　　　　(b)

perpendicular to the axis of the 300 Å fiber, instead of lying parallel to this axis, as in diagram (a).

Compute the dimensions **a–c** also for model (b); and compare them with the corresponding figures for model (a). (See Widom, J. and Klug, A. (1985), *Cell* **43**, 207–13, for evidence about key dimensions, and in particular that the axial spacing **c** is ≈ 110 Å.)

7.4 The telomeres of human chromosomes are made from a long, multiple repeat of the sequence (5') TTAGGG (3'), as stated in the caption of Fig. 7.6. Often, such a repeat projects beyond the end of the double-helical DNA at each end of the chromosome, to leave a short, single strand of 6 unpaired bases as shown below:

> (5') ... TTAGGGTTAGGGTTAGGG　　　(3')
> (3') ... AATCCCAATCCC　　　　　　　(5')

Such an arrangement is known as a "sticky end" of DNA. There are known to be many enzymes in the cell that can join or "fuse" two such single-stranded ends of DNA molecules to one another, if they detect sufficient Watson–Crick base-pairing.

a Could two identical (5') TTAGGG (3') ends like the one shown above (with the second one found by rotating the first through 180° in the plane of the diagram) be joined by such an enzyme?

b Suppose human telomeres were made not from repeats of TTAGGG, but from repeats of some other sequence such as TAGCTA or CGATCG. Would such hypothetical telomeres prevent the ends of chromosomes from joining to one another by Watson–Crick pairing? (See Ijdo, J.W.,

Baldini, A., Ward, D.C., Reeders, S.T. and Wells, R.A. (1991) *Proceedings of the National Academy of Sciences, USA* **88**, 9051–5, for a rare instance of telomere fusion.)

7.5 It is often possible to prepare, in the laboratory, a "soup" or extract of proteins from a living cell which is able to synthesize some particular RNA molecule when a DNA "template" molecule, having a specific sequence, is added to it.

If you study the process of transcription solely in this cell extract, on "naked", histone-free DNA, what important features of the chromosome structure—features that could affect the activity of a DNA template in a living cell—would be excluded from your study? By scanning the diagrams of this chapter, identify at least four different structures of DNA within a chromosome that might influence gene activity in this way. (See Felsenfeld, G. (1992) *Nature* **355**, 219–24, for a recent survey of such concerns.)

CHAPTER 8

Methods Used to Study the Structure of DNA

Our goal in this book has been to explain, as simply as possible, how DNA works in biology. For that reason, we have tried not to dwell too much on the methods which are used by scientists to study DNA: instead we have tried to give an integrated picture of DNA as obtained by many different methods of analysis. We have emphasized on many occasions that DNA is a very tiny object; yet our pictures of DNA have been drawn in terms of images which may be perceived by the reader at a "household" scale.

Now a student who wants to understand any subject in depth will want to know exactly how the evidence has been obtained, from which the overall conclusions have been reached. And the historian of whom we spoke in Chapter 1 was puzzled not so much by the fact that the DNA in the cells of our bodies is so exceedingly small, but by the problem of how one can *find out* anything about an object so small. In this chapter, therefore, we shall explain some of the techniques which scientists today use to study the structure of DNA.

The most important method, at least from a historical point of view, has been the analysis of DNA structure by *X-ray diffraction*. This is the tool which was used to discover the basic double-helical form of DNA in 1953. Ten years earlier, in the 1940s, studies of pneumococcal bacteria by Oswald Avery and colleagues had shown that a pure preparation of DNA could cause a harmless form of the bacterium to become infectious, and so impart pneumonia to mice. (We know now that the DNA used by Avery contained a gene for making a strong shell or coat around the bacterium, but this was not known at the time.) By the 1950s, enough evidence had piled up to convince even physical scientists that DNA might constitute the invisible "gene" of which biologists had spoken for more than 40 years.

Therefore, some physicists and chemists began to investigate the structure of DNA by various methods, including X-ray diffraction, to see whether its physical structure might shed any light on how DNA could act as the genetic material.

A distinguished early X-ray worker was Rosalind Franklin. She pulled fine fibers of DNA from natural sources, and found that when these fibers were exposed to X-rays, they could give either of two distinct X-ray diffraction photographs. She called these two patterns "A" and "B." The "A" form was seen when she kept the fibers relatively dry, while the "B" form was seen when she kept the fibers wet. Her "B" form photograph—which was much the simpler of the two—was interpreted by James Watson and Francis Crick in the spring of 1953 in terms of a right-handed double helix containing A–T and G–C base-pairs. Robert Langridge subsequently confirmed the essential points of the "B" form model in 1960; and Watson Fuller produced a similar model for the "A" form in 1965, refining a model first proposed by Franklin and Raymond Gosling late in 1953. These "A" and "B" form models were shown as part of Fig. 2.7.

Later work by Struther Arnott and colleagues showed that DNA of a regularly repeating sequence, such as A_n/T_n (that is, all A on one strand and all T on the other) or $(AT)_n/(AT)_n$ (that is, the alternating sequence ATATAT on both strands) could be extremely polymorphic. Thus, each fiber preparation could produce as many as three or four different kinds of X-ray pattern, depending on its salt and water content during exposure to X-rays. For example, a fiber of the sequence $(AT)_n/(AT)_n$ can produce a total of four different X-ray patterns under different conditions, which are known as "A," "B," "C," and "D." Other base sequences can produce X-ray patterns, for example "B'," "C'," and "C''," which are clearly variants of the basic forms; while still others show X-ray patterns such as "E" that are plain distinct. Thus, by 1980, much evidence had accumulated that the structure of DNA might be more complex than Watson and Crick could ever have anticipated in the 1950s. Yet although X-ray pictures of a fiber sample can show well enough that the forms "A" to "E" are distinct, they do not yield enough information to determine the detailed three-dimensional structures of these different forms at sufficient resolution to see the individual atoms clearly.

Fortunately, by 1980, chemists such as Keiichi Itakura, Shoji Tanaka, and Jacques van Boom had learned how to synthesize DNA chemically in large amounts, and how to purify it so that one could grow crystals of particular, short base sequences.

Crystals will not grow unless the preparation is pure; that is, unless the short fragments of DNA (or oligomers,[1] as they are known) have identical sequence and are all of the same length. The first structure of DNA to be solved by X-ray analysis of a crystal, as distinct from a fiber, was that of the sequence ATAT by M.A. Viswamitra in 1978. It proved to be disappointing, for the molecule did not form a complete double helix. But the next few X-ray structures were to produce astonishing results: the sequences CGCG and CGCGCG, as analyzed independently by Andrew Wang, J. Crawford, Horace Drew and their colleagues in 1979–80, each crystallized as left-handed double helices. Biologists had assumed for over 20 years that DNA could only be right-handed, and then it was discovered that DNA could be left-handed as well! Earlier solution studies by Fritz Pohl and Tom Jovin, using circular dichroism methods, had suggested that alternating C–G sequences such as CGCG might be either right-handed or left-handed, depending on the salt concentration; but only a few crystallographers and other specialists had taken them seriously.

Since 1980, many important structures of DNA oligomers have been analyzed in single crystals by X-ray diffraction methods. These studies form the essential background for all of the previous chapters in this book. They are too numerous to be cited here, but we list references to this large body of work at the end of the chapter. Here we shall explain the typical steps of an X-ray analysis with reference to one particular DNA molecule.

First, the crystallographer must decide what sequence of DNA to study, and then prepare large amounts of the material in pure form, usually by the method of chemical synthesis. Then, he or she must grow crystals of this substance that are suitable for X-ray diffraction studies. Growing crystals is a very chancy business. Some crystals of DNA, suitable for X-ray analysis, are shown in Fig. 8.1. They are each about 1 mm long. These particular crystals contain a complex of an eight-base-pair double helix of sequence AGCATGCT, together with the antibiotic nogalamycin to which it binds tightly. The antibiotic is orange; the DNA itself has no color. The crystals shown in this figure were grown slowly in a cold room for 2 weeks before they reached the required size of 1 mm.

Next, one of the crystals is carefully mounted in a wet, sealed capillary tube and placed in an X-ray beam. If the crystal is well-ordered in structure, an X-ray photograph such as that shown in Fig. 8.2 will be obtained. Any large crystal of DNA is

Figure 8.1 Crystals of DNA of sequence AGCATGCT in combination with the antibiotic nogalamycin. Each crystal is about 1 mm long. Courtesy of Maxine McCall and Louise Lockley.

Figure 8.2 Typical X-ray diffraction photograph of a DNA crystal. The DNA molecule in the crystal which produced this picture contained 12 base-pairs of sequence CGCGAATTCGCG on each strand.

made from millions of identical DNA molecules, all of which are close-packed into some sort of regular array. The geometrical locations of spots in Fig. 8.2 tell us what sort of repeating array the molecules have formed. DNA can pack into a crystal having any one of 65 different kinds of three-dimensional symmetry; and so the first task is to determine which kind of symmetry is present in any particular crystal. The next job is to measure the relative intensities of the spots, for each of many photographs taken with different orientations of the crystal in the X-ray beam. In the early days this was a time-consuming task, but now it can be done routinely in a short time by automated methods, and by use of a powerful X-ray beam. The final task, which is the most difficult, is to translate the relative intensities of spots into a model of the atomic structure, for DNA and any antibiotic or protein in the crystal. Each non-hydrogen atom (for example, carbon, nitrogen, oxygen, or phosphorus) can be located to an accuracy of about 0.1–0.2 Å in three-dimensional space if this last job is done properly; and so even the fine details of a structure can be found.

Figure 8.3 shows one small part of a completed DNA structure as determined in this way. There we can see an adenine–thymine base-pair, surrounded by many ordered water molecules. The

Figure 8.3 Assignment of atomic structure to part of an electron-density map as produced by X-ray diffraction methods. This part of the map shows an A–T base-pair as in Fig. 2.11(a). Water molecules are labeled "WAT." From Dickerson and Drew (1981) *Journal of Molecular Biology* **149**, 761–86.

locations of carbon, nitrogen, and oxygen atoms are identified by successive contours of increasing electron density. In fact, the X-ray scattering power of any atom is proportional to the square of its electron number; so carbon scatters X-rays as $(6)^2 = 36$, nitrogen as $(7)^2 = 49$, and oxygen as $(8)^2 = 64$. That is why hydrogen atoms cannot be located, because they scatter X-rays only as $(1)^2 = 1$. And that is also why heavy atoms such as bromine, iodine, or platinum can be used to help solve X-ray structures, because platinum, for example, scatters as $(78)^2 = 6084$, or much more strongly than the other, lighter atoms.

From the final assignment of atomic positions in a crystal, and after many cycles of refinement by a large computer, one can obtain a highly accurate model of the whole DNA molecule (plus protein or antibiotic, as appropriate) in three dimensions. Such three-dimensional models are usually regarded as being broadly representative of the structure in solution, on average: for if the structure in the crystal were too different from that in solution, the molecule would never have crystallized! These three-dimensional models form the whole underpinning for the science of molecular biology; and that is why we have explained how they are derived. In addition, the principles of symmetry and molecular structure which one learns during the course of an X-ray analysis are useful in understanding other, wider aspects of biology which do not have anything to do with crystals. For example, one cannot understand filaments of any sort, whether they take the form of flagella, muscles, microtubules, or DNA, without knowing something about symmetry. As J.D. Bernal once wrote, "generalized crystallography is the key to molecular biology."

We have not explained here how scientists actually convert the relative intensities in a set of X-ray patterns to a detailed three-dimensional atomic structure. Our reason for this is that the mathematics used in the process are extremely difficult; all the student needs to know is that the final structure is built up from the superposition of many waves, and that the relative intensity of each spot on the film defines the height of one particular wave. One should also know that the mathematics are highly statistical in nature: each X-ray particle (or photon) behaves unpredictably when going through the crystal, and can emerge at any spot. Thus, from a single X-ray scattering event you cannot learn anything, but by averaging over many events, you can obtain a consistent probability of the photon's arriving at any spot, which is then proportional to its intensity.

In addition to X-ray diffraction methods for analyzing atomic structure, one can also carry out *electron microscopy* experiments. Here one first lays a molecule of DNA onto a "grid," and applies a heavy-metal stain such as uranium or platinum in order to help visualize the DNA in the electron beam. Then, one surrounds the grid by a vacuum and shoots electrons through the sample. The electrons are focused, and clean pictures of DNA such as those shown in Figs 5.5 and 6.7 may be obtained. Electron microscopy experiments are easy to perform, but the pictures are lacking in atomic detail. Furthermore, the DNA— or whatever—can easily be distorted from its natural shape in the course of preparation for electron microscopy, since it must be removed from the fluid which normally surrounds it and be placed in a vacuum. The images produced by electron microscopy generally show the molecule of interest in just two dimensions, unless special care is taken to tilt the grid and shoot successive pictures from different perspectives.

Both X-ray structure analysis and electron microscopy are *direct* techniques for determining the structure of DNA. In the end, you can simply look at a three-dimensional model of DNA as determined by X-ray diffraction, or at a picture of DNA on a grid as determined by electron microscopy, and be confident that the thing you are looking at corresponds to physical reality. However, relatively few scientists practise these two methods, because: (a) the necessary equipment is expensive; (b) a scientist must be highly trained in order to carry out such analyses; and (c) it is often difficult to prepare a suitable crystal, or indeed a sample for electron microscopy, of a biologically interesting substance.

For these reasons, most scientists today employ a variety of *indirect* methods for finding out about the structure of DNA. Most of the indirect methods are less reliable than the direct methods described above, but they are generally cheaper and simpler to perform. We shall explain about several different kinds of indirect method here. First there are the *spectral methods* such as nuclear magnetic resonance, Raman spectroscopy and circular dichroism. Then there are the *enzymatic methods* such as "footprinting" with a DNA-cutting enzyme. Finally there are the *electrophoretic methods*, where DNA is passed through a gel in the presence of an electric field, and thereby separated according to its size, shape and electric charge.

In the technique of *nuclear magnetic resonance* or NMR, a concentrated sample of DNA is brought into the presence of a strong magnetic field, so that all of the hydrogen atoms in the DNA (which have magnetic moments) align themselves with this field. Then the sample is exposed to a low-energy electromagnetic field over a range of radiofrequencies; and individual hydrogens in the DNA absorb energy at some particular frequency so as to align themselves against the main field. The amount of energy required to flip a hydrogen atom against the main field is very sensitive to its location in the molecule, how it is bonded to other atoms, and what atoms are located near it in three-dimensional space. Figure 8.4 shows part of the NMR spectrum for a 12-base-pair molecule of sequence CGCGAATTCGCG at several different temperatures. Because both ends of the molecule are related by symmetry (in other words, CGCGAA can pair to TTCGCG), there are only six peaks rather than 12 in the spectrum. Each of these peaks represents the magnetic alignment of a single kind of hydrogen atom in millions of identical DNA molecules. These particular hydrogens lie in the center of Watson–Crick base-pairs as N–H. . .O or N–H. . .N hydrogen bonds (recall Fig. 2.11). There are six peaks, at slightly different frequencies of absorption, because there are six slightly different kinds of base-pair in different chemical environments. When the temperature of the sample is increased from −5°C to +51°C, the base-pairs on either end of the molecule (numbers 1 and 2) begin to fall apart: then the NMR peaks for their hydrogen atoms are lost, as they exchange with water. The virtue of NMR methods is that they provide information about the dynamic structure of DNA in solution, which is not available from X-ray or electron microscopy studies. Many of the applications of NMR to DNA in solution were pioneered by Dinshaw Patel at Bell Laboratories in the 1970s.

More recently, Kurt Wuthrich in Switzerland has applied a new technique in NMR spectroscopy, first developed by Richard Ernst, that allows you to measure the transfer of magnetic alignment from one hydrogen atom to any other in a chemical molecule, and then use this information to tell you something about its chemical or three-dimensional structure. The two hydrogens must be close together in space for such a transfer to occur. In principle, this technique might allow one to measure precisely all of the interatomic distances between different hydrogens in DNA, and so determine its precise structure in solution; but in practice the technique has proved only qualitatively useful. Thus, one can tell whether two hydrogens are close together or far apart in space, but not the precise distance that separates them. Such "transfer" methods are useful in studying complexes of DNA with antibiotics or with proteins, in order to see which hydrogens on the DNA are close to which

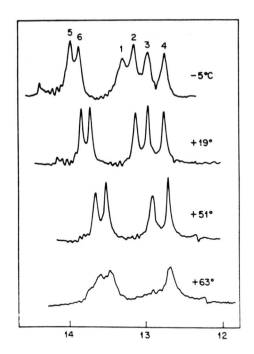

Figure 8.4 NMR spectra of a DNA molecule of sequence CGCGAATTCGCG, at four different temperatures. Courtesy of Dinshaw Patel.

hydrogens on the antibiotic or protein. NMR methods are limited to molecules having no more than a few thousand atoms, because of the increasing complexity of the spectra for many atoms, and because very large molecules do not turn over (or tumble) rapidly enough in solution to produce a clean spectrum.

Raman spectroscopy measures the vibrational frequencies of individual bonds in the DNA, and hence it is a sensitive measure of chemical bonding and structure. The best work using this method has been done by Yoshifumi Nishimura in Japan. He has calibrated his spectral lines for large DNA molecules in terms of the many different spectra seen for single nucleotides of known structure in a crystal. Nonetheless, such methods are limited in their power of structural interpretation, in comparison with X-ray diffraction or NMR. The main advantage of Raman studies is that they require very little material; thus a spectrum can be measured from any small part of a chromosome or a living cell.

Circular dichroism spectroscopy measures the absorption of polarized ultraviolet light by DNA, and shows whether it absorbs more left-handed or right-handed polarized light. This kind of result has no theoretical interpretation, however, and such studies have a long history of artefactual claims about DNA structure. Recently, Louise Fairall and Daniela Rhodes have calibrated many circular dichroism spectra of DNA in terms of the known structures of DNA, as seen in single crystals.

A variety of *enzymes* and *chemicals* can be used for the analysis of DNA sequence and structure. Some of these will break the DNA into bits at certain short series of nucleotides such as GAATTC, or else at certain single nucleotides such as A, G, C, or T. Others will cut the DNA in practically any location, except where an antibiotic or protein has bound itself to the molecule. Still others will cut the DNA only in places where they detect an unwound single strand, instead of a double helix. What all of these methods have in common is that they use *electrophoresis in gels* to separate the fragments of DNA according to their size. So we must say something about the motion of DNA through gels, before we can explain how enzymatic or chemical methods can be used to probe the structure.

A gel is nothing more than a three-dimensional array of tiny, randomly oriented fibers, like the fibers in a grass mat that you wipe your feet on before going into the house. Most of the spaces in a gel are filled with water. For example, when you make "jello" as a dessert at children's parties, you simply boil a small amount of gelatin powder in a large volume of water, and let it cool; then the final gel will be no more than 10% gelatin and 90% water. It is easy to see why a typical gel should be highly porous to small molecules such as DNA or protein: they can move easily through the gel by passing through the water between gel fibers.

Some small molecules can move only slowly through a gel by diffusion, because they are uncharged. But DNA and protein both carry a net electric charge, and so they can move quickly through a gel in the presence of an electric field. An ordinary gel as used in DNA work can be poured between two glass plates, or else as a thin slab onto a flat surface of size 20 cm × 20 cm, as shown in Fig. 8.5. Positively and negatively charged electrodes can be placed at its ends in a suitable salt solution in order to impart the desired voltage gradient.

We need not concern ourselves yet with details of the gel-running experiment, such as how to choose the correct density of gel; or indeed how to describe the motion of DNA through a gel by use of mathematics. For present purposes, only two things really matter: one is that short molecules of DNA can travel through a gel more rapidly than long ones, and the other

Figure 8.5 A typical gel-running apparatus. Other set-ups may be as long as 50 cm, and may be vertical as well as horizontal.

is that every kind of DNA molecule runs through the gel at a very definite, size-related speed.

Suppose that we have a pure sample of linear[1] DNA of some given length and sequence. If we load this sample into a small "well" at one end of the gel, and turn on the voltage for a few hours, we find that the DNA migrates as a tight band towards the other end, without significant broadening or diffusion. There are two commonly used ways to locate the DNA in a gel. One is to stain the gel with a dye such as ethidium bromide (see Chapter 2), which fluoresces strongly under ultraviolet light when it is bound to DNA. The other is to incorporate one or more radioactive phosphorus atoms into the DNA at its 5'- or 3'-end, or perhaps throughout the length of the molecule. Then the radioactive band of DNA will darken an ordinary photographic film after only a few hours.

Suppose next that we take the same sample of DNA, but treat it with a "restriction enzyme" that cuts wherever it can find the particular sequence of nucleotides GAATTC. If there are n sequences of the kind GAATTC along its length, our DNA sample will now run through the gel as a series of $n + 1$ distinct bands, as shown schematically in Fig. 8.6. This scheme enables us to find out how many sequences GAATTC are contained in the DNA, and something about their location, since the smaller fragments will run further down the gel than the larger ones.

Finally, suppose we treat the same double-stranded DNA with some chemical that cuts only at the nucleotide A on any strand. Then if the products of this chemical reaction are run through a "denaturing" gel that contains a high concentration of urea in order to separate the two strands, they will produce a "ladder" of single-stranded fragments that mark the relative locations of

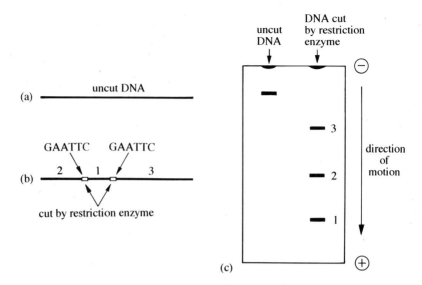

Figure 8.6 Running of DNA fragments through a gel, before and after cleavage with a restriction enzyme.

all bases A in the DNA sequence, as shown on the left-hand side of Fig. 8.7. In order to locate bases G, C, and T, one can use other chemical reactions that are specific for these bases, and run these DNA fragments through the gel as well (Fig. 8.7). You can select which strand of the double helix you want to see in the gel photograph, by attaching a radioactive phosphorus to either one strand or the other; fragments from the "other" strand will be non-radioactive and therefore invisible. Such methods can be used to determine the complete sequence of a DNA molecule as long as 400 nucleotides, because single strands of length 10–400 nucleotides all run at distinctly different speeds through a gel. Strands of length greater than about 400 nucleotides tend to run at more similar speeds, and they cannot easily be resolved from one another. In order to determine the complete sequence of a very long DNA, say from a chromosome or a virus, you have to break it into many different pieces of size 400 base-pairs or less, and then sequence the pieces individually.

That is not the only way of determining the sequence of a long DNA molecule, but it is representative of the other methods. The techniques described here have become so routine that scientists today are discussing whether they should try to determine the DNA sequence from all 23 pairs of human chromosomes!

All of the enzymes and chemicals discussed so far can cut the

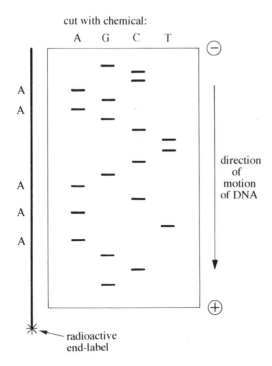

cut with chemical:

A   G   C   T

direction
of
motion
of DNA

radioactive
end-label

Figure 8.7 A gel for determining the sequence of a DNA molecule, which is shown in part on the left.

DNA in very precise locations, according to its base sequence. But certain other enzymes and chemicals, for example DNAase I, can cut the DNA in practically any location, with only a mild specificity for the base sequence. How might such a generalized DNA-cutting activity be useful?

Suppose we have isolated from an animal or plant some important protein that affects gene activity, by binding to an unknown DNA sequence along the length of a chromosome. How can we determine where it prefers to bind to the DNA? Usually the protein will bind so tightly to the DNA that it blocks the cutting activity of an enzyme such as DNAase I; so we can locate the bound protein by looking to see where the cutting activity of DNAase I is reduced in the presence of protein. This technique is known as "footprinting," because the regions of reduced cutting by DNAase I look like "footprints" of the protein on the DNA when we study a gel photograph.

One example of such an experiment is shown in Fig. 8.8. There we are looking to see where a small antibiotic called "echinomycin" binds along the DNA. Our detailed procedure is as follows: we label a DNA molecule of 150 base-pairs at either of its two 3'-ends with radioactive phosphorus atoms in separate experiments; then we add the antibiotic to each of these DNA preparations, and wait for a few minutes until the antibiotic has located its preferred binding sites; finally, we add DNAase I for a certain length of time, until some cutting has taken place at every nucleotide. When the two kinds of DNA sample are run on a denaturing gel, we obtain the patterns shown in Fig. 8.8.

The left- and right-hand sides of this figure show the results of DNAase I cutting along either of the two strands of the double helix. The first three gel lanes on either side show how

Figure 8.8 Use of a gel for "footprinting" of the antibiotic echinomycin onto a DNA molecule of 150 base-pairs. The left and right-hand sides of the figure show cutting by DNAase I on either of the two strands of the double helix, at different drug concentrations. Courtesy of Loretta Low and Michael Waring.

DNAase I cuts the free DNA, in the absence of echinomycin. The bands there show evidence for some cutting at every nucleotide; yet these bands are of greatly varying intensity. It seems that DNAase I prefers to cut more at some base sequences than at others. Many studies have shown that DNAase I binds across the minor groove of DNA, and only cuts well if this groove is of a correct size, and if the bond to be cut is positioned properly relative to the active site of the enzyme. Each of these structural features depends on the base sequence of the DNA, and so we see a rather complex pattern of cutting even in the absence of the antibiotic.

The effect of adding the antibiotic to DNA can be seen in the remaining six gel lanes on either side: different sets of lanes contain different concentrations of echinomycin. The lanes labeled "G" are markers for guanine. One can easily identify the sites of binding of echinomycin to DNA, by looking for short regions within any lane of the gel where the bands are relatively faint, as compared with their intensities for free DNA. The antibiotic binds tightly to seven different locations along this 150-base-pair DNA, and blocks DNAase I cutting for 5–6 bonds at each of its binding sites. "Site 1" is located near one end of the molecule, far from the radioactive end-label used on the left-hand side of the figure; so there it runs slowly, near the top of the gel. But "site 1" lies close to the radioactive end-label used on the right-hand side; so there it runs rapidly, near the bottom of the gel. The opposite holds true for "site 7," which runs off the bottom of the gel on the left, but near the top of the gel on the right. If we were to compare the locations of these echinomycin binding sites with the base sequence of the DNA used in this experiment, we would find that each binding site is centered on a short sequence of the kind CG. One year after this experiment was published, it was confirmed by X-ray diffraction methods that echinomycin binds at a step CG in its crystalline complex with DNA.

Near "site 2" on the left-hand side, or "site 1" on the right-hand side of Fig. 8.8, several bands adjacent to each echinomycin-binding-site actually become more intense in the presence of the antibiotic than in its absence. It turns out that these are AT-rich regions of high propeller twist, where DNAase I cuts only poorly in the free DNA. Once echinomycin binds next to such DNA, it flattens the propeller twist and lets DNAase I cut more rapidly. Other antibiotics such as ethidium bromide are also thought to flatten the propeller twist, so as to let DNAase I cut more rapidly.

Footprinting studies like the one shown can now be carried out even on single-copy DNA sequences within a whole chromosome, to see where different proteins bind along the DNA in living cells.

As a final example of useful enzymatic methods, let us look at the cutting activity of an enzyme called "S1 nuclease." This enzyme cuts DNA only where it can detect an unwound single-strand, as opposed to a double helix. Figure 8.9(a) shows two particular 20-nucleotide strands of DNA, called "20-Watson" and "20-Crick" respectively. When these two strands are mixed in equal amounts, they form a 20-base-pair double helix with Watson–Crick pairs as shown. But if the two strands are kept separate, each may fold back on itself to form a "hairpin loop," with six base-pairs in the stem and eight unpaired bases in the loop. These hairpin loop structures are shown schematically in Fig. 8.9(b).

When the 20-base-pair double helix from part (a) is treated with S1 nuclease, and run through a denaturing gel, the cutting pattern shown on the right-hand side of Fig. 8.9(c) is obtained. In this experiment, the 3'-end of 20-Watson has a radioactive label, while the strand 20-Crick remains unlabeled and therefore "invisible." It seems that S1 nuclease can cut only at the very ends of a double helix, where the two single strands of DNA are not so firmly connected to one another. Yet when you treat the hairpin 20-Watson from Fig. 8.9(b) with S1 nuclease, you see the pattern shown on the left-hand side of Fig. 8.9(c). Once again the lowermost part of the double helix is cut at its fraying end, but in addition you can see extensive cutting of the strand within its loop of eight unpaired bases, halfway up the gel. This result shows conclusively that S1 nuclease recognizes the structure of the DNA rather than its sequence, for the same sequence TTCCTCTT is not cut when it is part of a double helix on the right.

In summary, one can cut the DNA using a wide variety of enzymes and chemicals, in order to probe its structure. Some of these recognize the base sequence, while others recognize the double-helical structure, or lack thereof. In all such cases, the standard procedure is to separate the fragments of DNA so obtained by means of electrophoresis in gels, where each double helix or single strand runs according to its size.

Finally, although it is not widely known, it seems that the gels themselves can be used to find out something about DNA structure. It is not strictly true that molecules of DNA run in gels according to their *size*; rather, they run according to their

(a)   (b)

(c)

Figure 8.9 Investigation of hairpin-loops by means of cleavage with single-strand specific S1 nuclease, followed by gel electrophoresis of the cleavage products.

size, shape and *net electric charge*. For example, if a piece of negatively charged DNA is bound to a positively charged protein, it will have a reduced overall negative electric charge, and so it will run more slowly through a gel than free DNA of the same length. That is easy enough to understand; but how can a molecule of DNA run in gels according to its shape? How can the gel fibers sense the shape of the DNA, in addition to its chain length or size? This result also follows in a trivial way, once you understand the theory of DNA motion through gels. We shall explain briefly how it works, without recourse to the difficult mathematics that are required to account precisely for the motion of DNA through gels.

Actually, there are two kinds of gel electrophoresis, that use *constant* or *regularly alternating* electric fields. All of the examples which we have given so far have been for the method of constant field. In this case, a sample of DNA is applied to one end of a gel near a negatively charged electrode (see Fig. 8.5), and this sample then migrates through the gel for a distance of 20 cm (or 50 cm in some cases) towards a positively charged electrode. The potential difference between the two electrodes is typically 100–1000 V, or enough to generate a current of 10–100 mA for a typical salt solution, at neutral pH. Most of the current between the two electrodes is carried by positively and negatively charged ions in the salt solution, such as sodium (Na^+) and chloride (Cl^-), which move through the gel in opposite directions owing to their different charge. The DNA, of course, is negatively charged, so it moves in the same direction as the chloride ions; that is, it moves towards the positively charged electrode. It is necessary to keep a solution of salt in the pores of the gel, because DNA in distilled water will not move at all; it remains bound to sodium ions that pull it in one direction, while it wants to go in the other. In the presence of excess salt, DNA can move towards the positively charged electrode by exchanging sodium ions continuously as it moves forward.

When this experiment is carried out in a vessel filled with simple salt solution, in the absence of any gel, all DNA molecules move toward the positively charged electrode with about the same velocity, regardless of their size or shape. Thus, a single nucleotide will move at the same speed as several thousand base-pairs of double helix. The explanation for this is that the larger DNA molecules contain more phosphates, and hence they have more electric charge to pull themselves forward in the electric field; but they also experience more viscous drag from their contacts with water molecules, because they are bigger.

The two effects cancel, and so DNA moves through a simple salt solution at a speed independent of its size.

The purpose of a gel is to exert an even greater friction or drag on the larger DNA molecules, so as to cause them to move more slowly than the smaller ones. Gels can be made from fibers of agarose or polyacrylamide, at concentrations of 0.5–20 g of solid per 100 ml of fluid. Single nucleotides of DNA move through a gel at the same speed as they move free in solution, but large DNA molecules are retarded in proportion to their size by contacts with gel fibers. The gel method at constant field is capable of separating by size DNA molecules as long as 50 000 base-pairs, with a resolution of better than 1% the size of the DNA. It is also capable of separating by size a wide variety of protein molecules, with a similar resolution.

There is thus no mystery as to how a gel works, in principle. But how exactly does a gel sense the size and shape of the DNA (or a protein), and retard its motion accordingly? To answer this question, we must first see what a gel looks like on a molecular scale. A typical agarose gel, as visualized by electron microscopy, is shown in Fig. 8.10. It contains many fibers of diameter about 100 Å that cross over one another like the strands of a grass mat. The most important property of an agarose gel is that its fibers are arranged *randomly in space*. Suppose we could somehow make a gel from *regularly* arranged fibers or points, as shown schematically in Fig. 8.11(a). Then all of the passages through the gel would be of the same size, and hence a plot of particle velocity versus size would show a sharp cut-off where the particle (whether protein or DNA) becomes too large to go

Figure 8.10 Electron micrograph of a portion of a 2% agarose gel, 1 μm × 0.5 μm overall: small black rectangle is 1000 Å × 500 Å. Individual gel fibers are about 100 Å wide. Courtesy of Sue Whytock and John Finch.

through any passage. But, as we have said, real gels are constructed from fibers that are arranged randomly in space, as shown in Fig. 8.11(b). Thus some passages through the gel are small, while others are large, and the speed of the particle will decrease gradually with increasing size. Scientists usually adjust the concentration of their gel so that the size of the molecule they wish to isolate lies about in the middle of the smooth curve shown in the diagram.

How does the DNA pick its way through a series of randomly sized passages in a gel? Obviously, there can be no regular pattern to its motion, and so its path is usually described through the use of statistics. There are several ways of thinking about this. One way is shown in Fig. 8.12(a), where a "disc" must move from left-to-right across a plane field of randomly distributed point obstacles. In the free spaces it goes quickly, just like DNA free in solution; but where it overlaps one or more points it goes slowly, perhaps at only 5% or 10% of the

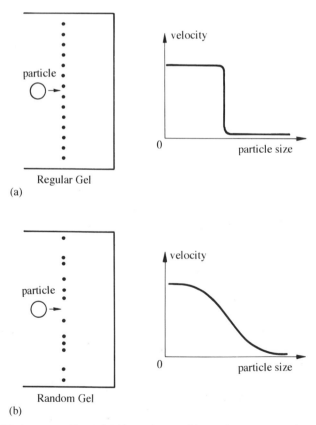

Figure 8.11 Sorting effect of (a) regular or (b) random-sieve gel.

speed of DNA in solution. The bigger the disc, the greater the fraction of its path where it contacts point obstacles: hence, big discs go more slowly through the field of obstacles than do small discs.

Another way of looking at the same thing is shown in Fig. 8.12(b). There we see cars traveling along a road, either singly or in caravans. The road is smooth for most of its length, permitting travel at 100 km/h, but it is rough in parts, permitting travel at just 10 km/h. A single car will average a speed of 92 km/h along this road, if the rough patches amount to just 1% of the total road surface: recall that average speed = (total distance)/(total time). But the long caravan will proceed at about 10 km/h, because it always contacts a rough spot somewhere along its length. Both parts of Fig. 8.12 describe a process of non-uniform motion, where the DNA starts and stops a lot while going through a gel. It does not proceed steadily, but its speed is uniform when averaged over a long period of time— which is itself short in comparison with the total time of testing.

The probability that a DNA molecule will make contact with a gel fiber is given by statistical theory for several different hypothetical cases. For a gel made of long fibers, the DNA is slowed according to its surface area; while for a gel whose obstacles are points, such as the junctions between fibers, the DNA is slowed according to its volume. Current experiments

(a)

(b)

Figure 8.12 Two analogies for gel electrophoresis of DNA at constant field: (a) a disc passes through a two-dimensional "gel" of randomly-located point obstacles; (b) a long caravan is impeded more than a single car by rough patches in a road.

suggest that DNA may possibly be slowed in a gel according to its volume, as if the gel behaved like a set of points arranged randomly in space.

Many experiments show that curved DNA moves more slowly through gels than does straight DNA of the same size or length. From the discussion presented in Chapter 5, it seems obvious that curved DNA will occupy a larger volume, or have a larger effective surface area, than straight DNA of the same size, because it can coil around itself to include a lot of empty space into which the gel fibers cannot enter. Thus, from careful measurements by gel electrophoresis, one can estimate the increased volume or surface area of curved DNA as compared to straight DNA. But one cannot learn anything definite about local structural parameters such as roll, slide or twist (see Chapter 3), because the gel does not sense such parameters directly.

Other experiments show that supercoiled DNA moves through gels with a velocity that depends on its linking number Lk. Some pictures of supercoiled DNA from bacteria were shown in Fig. 6.7. Circular DNA molecules with Lk near 0 move more slowly through gels than do molecules with a large value of Lk, either positive or negative, because the shape of the DNA becomes more compact as it gets more supercoiled. To be precise, it is thought that the cross-sectional diameter of an interwound supercoil decreases as Lk departs from zero. For example, in Fig. 6.5 the circle (a) with Lk = 0 has a larger cross-sectional diameter (or "fatness") than circles such as (c) to (e) with Lk = −3. Given a mixture of supercoiled DNA molecules of identical size but different linking numbers Lk, one can determine the true linking number Lk of any one of them by running mixture through an agarose gel, and then counting the number of discrete bands from the slowest with Lk = 0 at the top of the gel to, say, Lk = −10 or −20 near the bottom (see Exercise 8.6).

In general, the method of gel electrophoresis as described above can separate DNA molecules of different length out to about 50 000 base-pairs. Above this length the DNA is never out of contact with gel fibers: it is like a very long caravan in Fig. 8.12(b), always impeded by rough patches on the road. Thus, all DNA molecules beyond a certain length travel at the same speed; and so the gel is of little use in separating them.

To separate very large DNA molecules according to their size, one can carry out gel electrophoresis in the presence of a regularly alternating electric field. Suppose that the voltage which is applied to a gel is reversed at regular intervals of time:

how will that affect the motion of DNA through the gel? For example, the voltage from left to right in a gel may be set at +100 V for 10 s, then at −100 V for 5 s, then at +100 V again for 10 s, and so on. The DNA will move first in one direction for 10 s, then in the other direction for 5 s, then in the first direction again for 10 s. How will its net motion be altered? After an interval of 15 s, will it go simply $(10 − 5)/15 = 1/3$ as far as it would have gone at a constant 100 V? If the DNA is small, say less than 1000 base-pairs, then periodic reversal of the voltage has precisely this effect: the DNA goes 1/3 as far as it would have gone at constant voltage. But if the DNA is very large, say 100 000 to 10 million base-pairs, then periodic reversal of the field has a dramatic and unexpected effect: the large DNA cannot "turn around" in the gel as quickly as small DNA, when the voltage is reversed, and so the large DNA proceeds much more slowly than expected. In other words, there is a "time delay" for the large DNA to change direction in the gel; and consequently it does not go as far through the gel as expected. In the absence of this effect, all of the very large DNA molecules would go through a gel at the same speed.

One possible mechanism of this time-delay is shown schematically in Fig. 8.13; but no precise theory for this effect has yet been formulated. During the first 10 s at +100 V, the DNA proceeds from left to right at the expected speed. Studies by light microscopy show that the long DNA is sperm-shaped, with most of its mass concentrated into a "head," while a small part follows as a "tail." The large head of DNA must force its passage between gel fibers, like an icebreaker or bulldozer, because there are few if any pre-existing passages in the gel large enough to accommodate it. This kind of motion of DNA through the gel does not depend much on DNA size. But when the voltage is reversed, to −100 V, the DNA must reorganize itself to create a new "head" in the opposing direction. For a brief interval of time, the DNA is suspended over the gel fibers in motionless equilibrium, like a tie over a coat-hook. After this brief interval or "time-delay," the DNA falls from its unstable position and forms a "head" that can proceed in the reverse direction. The time required for the DNA to "fall off" the coat-hook depends quite strongly on DNA size. The same time-delay will occur when the voltage is switched back to +100 V. This interesting and unexpected behavior of large DNA in gels was discovered by David Schwartz and Charles Cantor in 1984. It enables one to separate in size, by means of gel electrophoresis, DNA to a length of 5 or 10 million base-pairs. Thus, one can separate in a gel all the chromosomal DNA molecules from a simple organism such as yeast, where the chromosomes do not exceed 10 million base-pairs. But one cannot yet separate by this method the chromosomal DNA molecules from a human being, which are of the order of 50–400 million base-pairs in length.

The key to the separation of large DNA molecules in gels, by this method, is to adjust the times for the forward and reverse pulses of voltage to be of the same order as the time to "turn around," for any given length of DNA. To a first approximation, the time required for a DNA molecule to "turn around" in the gel increases as the 2/3 power of its size. This is not hard to understand in principle: short ties fall from coat-hooks much more frequently than long ties.

One can also understand why DNA larger than 5–10 million base-pairs cannot be separated in size: when a tie gets very long, say the distance from your floor to your ceiling, then the time required for it to fall from a coat-hook, by random vibration, will not depend very much on its length. Often one can improve the separation of such DNA in gels by fiddling with the voltages,

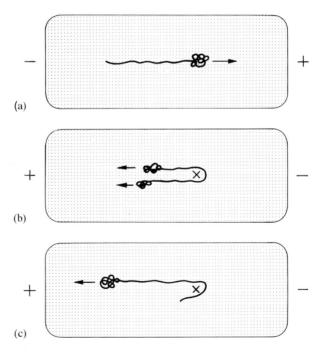

(a)

(b)

(c)

Figure 8.13 Movement of a very long DNA molecule through a gel, in alternating-field electrophoresis: (a) forward motion; (b) "tie on coathook" situation; (c) reverse motion.

or by reversing the voltage at an angle slightly less than 180° (say 120°); but the fundamental difficulty of separating 50–400 million-base-pair DNA in a gel remains. Whoever can solve this problem will be a great hero to molecular biologists, because he or she will for the first time make possible the detailed study and manipulation of human chromosomes. There seem to be two possible paths to a solution of this problem: (a) to change the structure of the gel so that its "coat-hook" properties apply to larger DNA; or (b) to reduce the contour length of the DNA uniformly, perhaps by wrapping it around proteins, to make the DNA seem smaller to the structure of the gel. Here is an interesting problem for a clever student.

Note

1. See Appendix 1.

Further Reading

Calladine, C.R., Collis, C.M., Drew, H.R., and Mott, M.R. (1991) A study of electrophoretic mobility of DNA in agarose and polyacrylamide gels. *Journal of Molecular Biology* **221**, 981–1005. Application of the ideas of Ogston to DNA gel-running. The source of the photographs in Fig. 5.5.

Diekmann, S. and Wang, J.C. (1985) On the sequence determinants and flexibility of the kinetoplast DNA fragment with abnormal gel electrophoretic mobilities. *Journal of Molecular Biology* **186**, 1–11. The kinetoplast DNA curves naturally in a plane, giving Wr = 0; but gives non-zero Wr when distorted by supercoiling, due to its curvature of about 1.3° per base-pair.

Drew, H.R. (1984) Structural specificities of five commonly-used DNA nucleases. *Journal of Molecular Biology* **176**, 535–57. Source of the S1 nuclease gel shown in Fig. 8.9.

Hagerman, P.J. (1985) Sequence-dependence of the curvature of DNA: a test of the phasing hypothesis. *Biochemistry* **24**, 7033–7. First proof that the slow gel-running of certain DNA sequences is due to their curvature; and a deduction that these sequences proceed through the gel as superhelices of definite shape.

Heller, C. and Pohl, F.M. (1989) A systematic study of field-inversion gel electrophoresis. *Nucleic Acids Research* **17**, 5989–6003. How very large DNA moves in gels with respect to its size and pulse times.

Keller, W. (1975) Determination of the number of superhelical turns in simian virus 40 DNA by gel electrophoresis. *Proceedings of the National Academy of Sciences, USA* **72**, 4876–80. Gel-running of supercoiled DNA in the absence or presence of ethidium bromide, an intercalating dye.

Kennard, O. and Hunter, W.N. (1989) Oligonucleotide structure: a decade of results from single-crystal X-ray diffraction studies. *Quarterly Reviews of Biophysics* **22**, 327–79. A detailed summary of X-ray work on DNA in single crystals.

Low, C.M.L., Drew, H.R., and Waring, M.J. (1984) Sequence-specific binding of echinomycin to DNA: evidence for conformational changes affecting flanking sequences. *Nucleic Acids Research* **12**, 4865–79. Source of the footprinting gel shown in Fig. 8.8.

Marini, J.C., Levene, S.D., Crothers, D.M., and Englund, P.T. (1982) Bent helical structure in kinetoplast DNA. *Proceedings of the National Academy of Sciences, USA* **79**, 7664–8. First indication that the slow gel-running of certain DNA sequences might be because they are curved.

Ogston, A.G. (1958) The spaces in a uniform random suspension of fibers. *Transactions of the Faraday Society* **54**, 1754–7. The classic work on the application of random statistics to gel-running phenomena, on which almost all later studies are based.

Rodbard, D. and Chrambach, A. (1971) Estimation of molecular radius, free mobility and valence using polyacrylamide gel electrophoresis. *Analytical Biochemistry* **40**, 95–134. Application of Ogston's work to protein gel-running.

Smith, S.B., Heller, C., and Bustamente, C. (1991) Model and computer simulations of the motion of DNA molecules during pulsed-field gel electrophoresis. *Biochemistry* **30**, 5264–74. The best theoretical study so far of alternating-field electrophoresis.

Steitz, T.A. (1990) Structural studies of protein–nucleic acid interaction: the sources of sequence-specific binding. *Quarterly Reviews of Biophysics* **23**, 205–80. A survey of how proteins bind to DNA in crystals.

Whytock, S. and Finch, J.T. (1991) Substructure of agarose gels as prepared for electrophoresis. *Biopolymers* **31**, 1025–8. Source of Fig. 8.10.

Bibliography

Avery, O.T., MacLeod, C.M., and McCarty, M. (1944) Induction of transformation by a desoxyribonucleic acid fraction isolated from *Pneumococcus* Type III. *Journal of Experimental Medicine* **79**, 137–58. The classic work implicating DNA as the substance of genes.

Bernal, J.D. (1967). *The Origin of Life*. Weidenfeld & Nicolson, London.

Crawford, J.L., Kolpak, F.J., Wang, A.H.-J., Quigley, G.J., van Boom, J.H., van der Marel, G., and Rich, A. (1980) The tetramer d(CpGpCpG) crystallises as a left-handed double helix. *Proceedings of the National Academy of Sciences, USA* **77**, 4016–20. The sequence CGCG in a low-salt crystal form.

Drew, H.R., Takano, T., Tanaka, S., Itakura, K., and Dickerson, R.E. (1980) High-salt d(CpGpCpG), a left-handed Z' DNA double helix. *Nature* **286**, 567–73. The sequence CGCG also goes left-handed in a high-salt crystal form.

Fairall, L., Martin, S., and Rhodes, D. (1989) The DNA binding-site of *Xenopus* transcription factor IIIA has a non-B form structure. *EMBO Journal* **8**, 1809–17. Calibration of the circular dichroism spectra of DNA between crystal and solution.

Franklin, R.E. and Gosling, R.G. (1953) Structure of sodium thymonucle- ate fibres: importance of water content. *Acta Crystallographica* **6**, 673–7. First identification of the "B" and "A" types of DNA in fibers.

Fuller, W., Wilkins, M.H.F., Wilson, H.R., Hamilton, L.D., and Arnott, S. (1965) The molecular configuration of deoxyribonucleic acid: X- ray diffraction study of the "A" form. *Journal of Molecular Biology* **12**, 60–80. First detailed model for the "A" form of DNA.

Itakura, K. and Riggs, A.D. (1980) Chemical DNA synthesis and recombinant DNA studies. *Science* **209**, 1401–5. A survey of the advances made in biology by the chemical synthesis of DNA.

Langridge, R., Marvin, D.A., Seeds, W.E., Wilson, H.R., Hooper, C.W., Wilkins, M.H.F., and Hamilton, L.D. (1960) The molecular configuration of deoxyribonucleic acid: molecular models and their Fourier transforms. *Journal of Molecular Biology* **2**, 38–64. First detailed model for the "B" form of DNA.

Leslie, A.G.W., Arnott, S., Chandrasekaran, R., and Ratliff, R.L. (1980) Polymorphism of DNA double helices. *Journal of Molecular Biology* **143**, 49–72. The wide variety of double-helical forms "A, B, C, D, E" and their variants as seen by fiber X-ray diffraction.

Nishimura, Y. and Tsuboi, M. (1986) Local conformations and polymor- phisms of DNA duplexes as revealed by their Raman spectra. In *Spectroscopy of Biological Systems* (eds R.J.H. Clarke and R.E. Hester), ch. 4, pp. 177–232. John Wiley & Sons, New York. Probing fine details of DNA structure by use of Raman spectra.

Patel, D.J., Pardi, A., and Itakura, K. (1982) DNA conformation, dynamics and interactions in solution. *Science* **216**, 581–90. A review of some of the earliest work by NMR on large synthetic DNA. Source of parts of Fig. 8.4.

Pohl, F.M. and Jovin, T.M. (1972) Salt-induced cooperative confor- mational change of a synthetic DNA: equilibrium and kinetic studies with poly (dG-dC). *Journal of Molecular Biology* **67**, 375–96. The left- to-right transition of DNA in solution.

Schwarz, D.C. and Cantor, C.R. (1984) Separation of yeast chromosomal-sized DNAs by pulsed-field gradient gel electrophoresis. *Cell* **37**, 67–75. The discovery that alternating electric fields can separate very long DNA in gels according to size.

Viswamitra, M.A., Kennard, O., Jones, P.G., Sheldrick, G.M., Salisbury, S., Falvello, L., and Shakked, Z. (1978) DNA double-helical fragment at atomic resolution. *Nature* **273**, 687–8. The crystal structure of ATAT, which "melts" at the central TA step.

Wang, A.H.-J., Quigley, G.J., Kolpak, F., Crawford, J.L., van Boom, J.H., van der Marel, G., and Rich, A. (1979) Molecular structure of a left-handed double-helical DNA fragment at atomic resolution. *Nature* **282**, 680–6. Discovery of left-handed DNA in a sequence CGCGCG.

Wuthrich, K. (1989) Protein structure determination in solution by nuclear magnetic resonance spectroscopy. *Science* **243**, 45–50. Modern state-of-the-art methods for analyzing protein and DNA structures in solution.

Exercises

8.1 In an electron-density map such as that shown in Fig. 8.3, which of the atoms should scatter X-rays most strongly?

   a   Rank in order of scattering power, from highest to lowest: carbon, phosphorus, oxygen, hydrogen, nitrogen.

   b   Which part of the electron-density map in Fig. 8.3 was produced by the strongest scattering of X-rays? Use Figs 2.8(b) and 2.11(a) to identify the atom types.

   c   Why are the hydrogen atoms in the Watson–Crick base-pairs not seen in Fig. 8.3?

   d   Which parts of the DNA might appear most strongly in a low-resolution X-ray diffraction structure such as that shown in Fig. 7.2?

8.2 Referring to Fig. 8.4, how many NMR peaks would you expect to find in the spectral region shown, and at $-5°C$, for double-helical DNA molecules which are specified by the following single-strand sequences?

   a   (5′) CGCAATTGCG (3′)

   b   (5′) AGCATGCATGCT (3′)

   c   (5′) AGCATGCGCG (3′)

In each case, first construct the two-stranded version of the molecule, and then look for two-fold symmetry that may make some peaks equivalent to others.

8.3 Suppose we have a DNA molecule of length 1000 base-pairs, which contains three sites of cutting, GAATTC, for a particular restriction enzyme; and suppose that these sites are located at 100, 350 and 550 base-pairs, respectively, from one end.

a On complete digestion of the DNA by the restriction enzyme, how many kinds of DNA fragment will be produced, and of what size?

b Which fragment will run fastest in an ordinary electrophoretic gel, and which will run slowest?

8.4 a Suppose that you have the DNA from a virus, consisting of 100 000 base-pairs, and you wish to determine its complete nucleotide sequence. Find, approximately, the smallest number of fragments which you could sequence individually, by use of present-day gel technology, in order to complete this task. Don't count the many overlapping fragments which would be needed to align partial sequences.

b A typical human cell contains 6×10^9 base-pairs of DNA, located on 23 pairs of chromosomes. What is the smallest number of fragments which you could sequence individually, in order to carry out the gigantic task of sequencing the complete DNA from a human cell? Again, don't count overlaps.

8.5 Figure 8.12(a) shows a flat disc migrating across a plane containing randomly spaced point obstacles. A more realistic model for gel electrophoresis would have circular disc obstacles instead of points. But it turns out that the second model is equivalent to the first, for practical purposes, if the radius of the moving disc or DNA is increased by the radius of the disc obstacles (or gel fibers).

Now, when long, superhelically curved DNA runs through a gel, it migrates as if its volume were that of an imaginary cylinder, circumscribed around the outermost parts of the curved DNA, rather than that of the DNA itself. Taking the radius of the DNA as 10 Å, and allowing an additional 50 Å for the radius of the gel fibers (which may be added onto that of the DNA, to give a combined radius of 60 Å for a straight DNA cylinder), calculate the ratio of the "circumscribed volume" to the "true volume" for the curved DNA $(A_6N_4)_n$ shown in the electron micrographs of Fig. 5.5. As structural parameters for the curved DNA use: $N^* = 178$ bp, $r = 20$ bp, $p = 128$ bp (see text of Chapter 5, p. 102), and assume here that 1 bp $= 3.3$ Å.

8.6 As explained in the text, closed circular DNA molecules run at a variety of speeds in a gel. That is because different linking numbers Lk can produce different interwound shapes, as shown, for example, in Figs 6.4(e) and 6.5(e) for Lk = ±3 or in Figs 6.4(a) and 6.5(a) for Lk = 0. Now it turns out that specimens with an open circular form (Wr = 0) run most slowly through a gel, while specimens with higher (whether positive or negative) values of Wr run more rapidly, since they are more compact and so present a smaller effective volume to the interfering gel fibers which retard forward motion.

This pattern of behavior enables one to pick out specimens having Lk = 0 from a gel; but one cannot easily pick out a specimen having, say, Lk = +3 from one with Lk = −3, because they both run at very similar speeds. Also, one cannot easily distinguish large values of Lk, say +19 versus +20, from each other, because these molecules are not sufficiently different in shape to run at significantly different speeds.

Scientists get around these problems by a technique which involves the addition of some ethidium bromide to the gel, so that in any given specimen the DNA untwists (see Exercise 2.2), thereby acquiring a more positive value of Wr, since Lk does not change.

Given that the addition of a certain amount of ethidium bromide imparts exactly Tw = −12 to all molecules in a sample, predict the values of Lk and Wr for the slowest-running specimen in the range Lk = 0 to −20. Which will be the fastest-running specimen?

(Use the equation on p. 121, and assume that the slowest molecule always has Wr = 0, while the fastest molecule has maximal Wr, whether positive or negative.)

Postscript

Our book is a small one, but the investigation of DNA is a large subject, and potentially very large indeed. We have included eight chapters on what we consider to be the most fundamental and well-understood aspects of DNA research, but we have omitted several potential chapters that might have dealt with other, less well-understood subjects such as: (a) the chemical theory of base-stacking energies in DNA, (b) the recognition of specific DNA sequences by proteins, (c) methylation of the DNA bases and its implications for biology, and (d) recombination or "restructuring" of the DNA molecule during cell growth.

The chemical theory of base-stacking energies in DNA is discussed briefly in Appendix 2, with reference to the recent work of Chris Hunter and Jeremy Sanders. In a few years' time, we expect that many aspects of DNA structure will be understood theoretically, in terms of the well-established principles of chemistry; and then these energies can be discussed in greater detail.

The recognition of specific DNA sequences by proteins was mentioned briefly in Chapter 4. There we said that proteins recognize DNA according to a two-step process: first they "dock" to an overall configuration of the sugar–phosphate chains that may depend on the base sequence; and then they "probe" for individual atoms on the exposed edges of base pairs. Much more could be said about each of these steps. For example, one could explain how roll, slide, twist and propeller twist, on a local scale, change the outward appearance of the sugar–phosphate chains so as to affect the "docking." Also, one could provide many examples of "probing" as seen in single crystals of proteins with DNA. The chemical theory of base-stacking energies has strong implications for the recognition of DNA by proteins, because those partial charges on the base rings, that influence how the bases stack on one another, are almost as large as the partial-charge or "hydrogen-bonded" interactions of DNA with a protein. In a few years' time, the interactions between protein and DNA should be better understood.

Some people regard the methylation of DNA bases—for example, the addition of a methyl group (CH_3) to cytosine in the same place where a methyl group is found in thymine—to be one of the most promising avenues for understanding the cellular development of higher organisms. Some enzymes add these methyl groups to DNA, while others take them off; and it is well-known that small changes in the pattern of methylation along any given segment of DNA can have large effects on the activity of genes nearby, as well as on the local chromosome structure. In a few years' time, we may know more about the role of DNA methylation in biology.

Finally, it should be emphasized that the order of base-pairs along a DNA molecule is not rigidly fixed, but can change during cell development, and before the fertilization of an egg cell by a sperm, when homologous pairs of chromosomes join together (in duplicate) to make the precursors to sex cells. These changes are known in general as "recombination" because a variety of enzymes can cause the DNA to recombine or restructure itself. Such recombination events are very important as a tool in genetics, for mapping the close proximity of genes along a chromosome; and they have also been considered as one means of fixing genetic defects in humans: one might insert a new piece of DNA into the human chromosome, so as to replace a faulty gene by a good gene. The mechanisms of recombination are not yet well-understood, but rapid progress is now being made.

Whether we consider base-stacking effects, or protein–DNA interactions, or methylation, or recombination, there is one overriding feature that distinguishes a well-understood field of science from a poorly understood one. In a well-understood field, there is a well-developed sense of the relations between things, and how they are structured relative to one another; often a single theory may explain hundreds of experimental observations. By contrast, in a poorly-understood field there are an abundance of "effects" and "factors" and long, hard-to-remember words, but few relations among them in a theoretical sense.

Our goal, therefore, should be to distinguish what is known from what is not known, and not ever to pretend that word-knowledge is a substitute for structural-knowledge. This important idea was expressed clearly long ago by John

Locke (1690), as a comment on his *Essay Concerning Human Understanding*:

> We cannot but think that angels of all kinds much exceed us in knowledge, and possibly we are apt sometimes to envy them that advantage, or at least to repine[1] that we do not partake with them in a greater share of it. Whoever thinks of the elevation of their knowledge above ours, cannot imagine it lies in a *playing with words*, but in the contemplation of things, and having true notions about them; a perception of their habitudes and relations one to another. If this be so, methinks we should be ambitious to come in this part, which is a great deal in our power, as near them as we can.

Note

1. Repine: to fret, be discontented.

Further Reading

Davidson, S., Crowther, P., Radley, J., and Woodcock, D.G. (1992) Cytotoxicity of 5-aza-2′-deoxycytidine in a mammalian cell system. *European Journal of Cancer* **28**, 362–8. Loss of DNA methylation causes metaphase chromosomes to assume a highly elongated shape, like that seen for the yeast part of the yeast–mouse chromosome shown in Fig. 7.7(c).

Holliday, R. (1989) A different kind of inheritance. *Scientific American* **260**, June, 60–73. A discussion of how DNA methylation affects gene activity.

Stahl, F.W. (1987) Genetic recombination. *Scientific American* **256**, February, 52–63. A good review on several aspects of recombination.

Appendix 1: Notes on the Derivation of Some Technical Terms

All of the terms in the following list have been introduced and explained in the text. The sole point of these notes is to explain the philological background of the words.

Cell. A term first used by Robert Hooke in 1665, on seeing the structure of sliced cork in a microscope. The structure was similar to that of cells in a honeycomb, monastery or prison.

Circular. Like a circle in the sense of being endless: not necessarily even a plane figure (cf. linear).

Chromosome. Filamentous body in the cell nucleus which appears colored (Greek: *chromos* = color, *soma* = body) in the light microscope after treatment with chemical stains.

Enzyme. An organic catalyst, usually a protein. From the Greek *zumosis* = leaven, used to describe the process of fermentation.

Gene. From the Greek *gen*, to produce. Cf. generator, generation, Genesis.

Homologous. Corresponding. From the Greek: *homo* = same, *logos* = word.

Interphase. From two Greek words: *phasis* = appearance and *inter* = between; used to describe an appearance in the middle of the cell cycle.

Linear. Like a line in the sense of having two ends and no branches: not necessarily straight, or even planar (cf. circular).

Metaphase. From two Greek words: *phasis* = appearance and *meta* = beyond or after; used to describe an appearance near the end of the cell cycle.

Nucleus. From the Latin *nux* = a nut; the central kernel of the cell.

Oligomer. A short example (Greek: *oligo* = few) of a long-chain molecule.

Polymorphic. Having many shapes. From two Greek words: *poly* = many, *morphe* = form.

Protein. From the Greek *proteios* = of the first rank. First applied to the large group of compounds by Muldner, acting on a suggestion made in 1838 by the famous chemist Berzelius.

Topoisomerase. From the Greek *topos* = a place and *isos* = equal. The ending -ase denotes an enzyme. An enzyme that changes the linking number Lk of DNA, to make different forms of DNA equal.

Toroid. A ring of the kind put through the nose of a bull: Latin *taurus* = bull.

Translation. From the Latin: *trans* = across, *latum* = past participle of "bring." Hence, movement of a body without rotation.

Appendix 2: The Chemical Theory of Base-stacking Interactions in DNA

In Chapter 2 we explained how the water-repelling hydrophobic quality of the bases could push a "ladder" of DNA into a twisted, double-helical shape; and we called this a first-order effect on the structure of DNA. Then in Chapter 3 we invoked the same hydrophobic property in order to account for the "propeller twist" which is usually found in the Watson–Crick base-pairs, and we described this as a second-order effect. Propeller twist makes the stacking of base-pairs onto one another much less straightforward than for planar base-pairs, and it thereby provides a key to understanding the conformational behavior of double-helical DNA. For example, propeller twist together with the difference in size between the large purines (G, A) and the small pyrimidines (C, T) enables us to understand the two different types of stacking which can occur at a pyrimidine–purine step; and this opens the way to an understanding of the classical "B" and "A" forms of DNA, as described in Chapter 3.

Thus, throughout the book, we have represented the bases and base-pairs as simple rectangular blocks, which stack onto each other with as much overlap as possible, on account of the water-repelling property of their flat surfaces. This sort of model obviously constitutes a simplification of the actual chemical forces that are involved in the stacking of bases on one another; but it is somewhat justified because it provides us with a clear explanation of the double-helical structure of DNA, and of propeller twist in the base-pairs.

In order to explain more complex phenomena, however, this level of simplification is rather too crude; and so we need a more detailed theory for the stacking interactions between adjacent bases. A new chemical theory of this kind has recently been proposed by Chris Hunter and Jeremy Sanders. We shall give here a sketch of this theory, and then show how it can explain some well-known observations concerning the behavior of DNA in fibers and crystals. The theory was originally developed in relation to the behavior of some large, flat molecules called "porphyrins" that are used by the body for many different purposes, such as carrying electrons in the mitochondria and carrying oxygen in the blood. But the same theory is also very helpful in the study of DNA.

Hunter and Sanders say that there are three general kinds of electrical force that control the interactions between molecules such as porphyrins, or the base-pairs of DNA. The first force is the one we have described already, namely a tendency for the base-pairs to stack on one another as fully as possible, in order to escape from contact with water. Why should this be so in molecular terms? As shown schematically in Fig. A1, water molecules in solution have a strong tendency to associate with other water molecules by means of hydrogen-to-oxygen or H ... O bonds, which are electrical in nature. Normally any water molecule will have four partners: two that place their hydrogen atoms in contact with the water oxygen, and two that place their oxygen atoms in contact with each of the water hydrogens. This intricate network of hydrogen bonding between water molecules is responsible for the "stickiness" of water, for example its tendency to form drops when falling slowly out of a tap. But when these water molecules contact an oily, water-insoluble DNA base, many hydrogen bonds of the kind

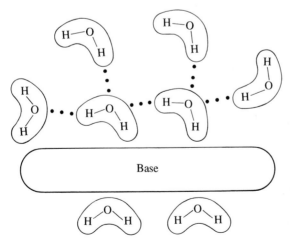

Figure A1 Schematic diagram showing how H ... O bonds between water molecules are disrupted by a hydrophobic base. The loss of such hydrogen bonds can be minimized when two base-pairs stack directly on top of one another, as in Figs 2.4 and 2.5.

H . . . O are lost, as shown in the diagram. Hence, the optimal arrangement of base-pairs in water is as shown previously in Figs 2.4 and 2.5, where the bases stack on one another as completely as possible in order to exclude water. In a solution of water mixed with ethanol (as in strong whisky) or, for example, at low water content in fibers of DNA, this kind of water-exclusion force becomes rather weak, so the base-pairs no longer need to stack as firmly on one another.

The second kind of force between base-pairs comes about because the upper and lower surfaces of any base (or base-pair) have a slight negative charge. As shown in Fig. A2(a), the negatively charged electrons that make up the "aromatic" system of bonding between atoms in any base lie mainly above and below the main part of the ring. Yet the positively charged nuclei of the various atoms that join to form a base (such as carbon, nitrogen and oxygen) lie near the center of the ring, when viewed edge-on, as in Fig. A2(a). So the overall distribution of electric charge in any base looks somewhat like a "sandwich," where the electrons are the bread and the nuclei are the cheese.

Now, when two bases (or base-pairs) contact one another along their upper and lower surfaces, they must repel one another to some extent, because of a negative-to-negative charge repulsion. An idealized, vertical stacking of two base-pairs is shown in the center part of Fig. A2(b) for the case of slide $S = 0$. We said previously that a fully stacked geometry is favorable because it excludes water; but now we see that it is

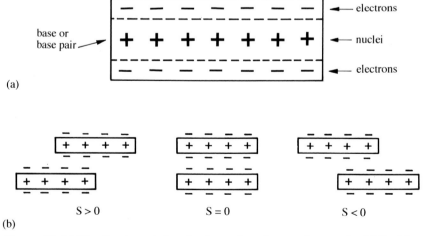

Figure A2 (a) Distribution of electric charge in a base or base-pair. (b) Positive or negative slide at a base-pair step reduces the repulsion between the negative charges, and produces some attraction between positive and negative charges in different base-pairs.

highly unfavorable when viewed in the context of our second force, which says that there is a strong negative-to-negative repulsion between base-pairs in a fully stacked geometry. Were it not for the water, then the two base-pairs would repel one another like the North poles of two magnets. Generally, the base-pairs in DNA will slide away from one another in a left-to-right sense to escape this repulsion, because the flexibility of sugar–phosphate chains is greatest in that direction. Thus, as shown in Fig. A2(b), the base-pairs can lie offset from one another at either positive slide S, as shown on the left-hand side of the drawing, or at negative slide S, as shown on the right-hand side. In either case, the negative-to-negative charge repulsion is lessened by the increased distance between pairs, and some positive-to-negative attraction may come about between nuclei in one base-pair and electrons in the other.

The "bread-and-cheese" effect is the exact opposite of the "hydrophobic" effect, so far as the stacking of adjacent base-pairs on one another is concerned. Which of the two is the stronger? This depends critically on the amount of water which surrounds the bases. When conditions are wet, the hydrophobic effect wins out; but when conditions are dry, the bread-and-cheese effect is stronger.

These notions explain rather simply the structural behavior of double-helical DNA in fibers. Originally, in the 1950s, it was not possible to grow crystals of DNA, because scientists had not yet learned how to synthesize DNA chemically; so most scientists had to isolate chromosomal DNA from natural sources such as calf thymus or salmon sperm, and then pull it out into long fibers, in order to study its structure by X-ray diffraction (see Chapter 8). When these fibers were wet, the X-ray photographs showed a "B" form structure with 10 base-pairs in a complete helical turn, and with base-pairs stacked vertically over one another at slide $S = 0$. But when the fibers were dry, either an "A" form with 11 base-pairs per turn, or a "C" form with 9.0 to 9.3 base-pairs per turn were observed. In the "A" form, the base-pairs stack offset at a positive slide $S = +1.5$ Å, while in the "C" form they stack offset at a negative slide $S = -1.0$ Å. Note that slide S and twist T change in tandem, as described in Fig. 3.13.

Not all possible forms of double-helical DNA can be seen in fibers, because the packing of long, thin DNA molecules into fibrous bundles often restricts their helical repeat to integral values such as 10 or 11 base-pairs per turn. Yet when such structures are observed, we can understand their behavior as follows: when the fiber is wet the hydrophobic forces are dominant, so the base-pair stacking is one of low slide near $S = 0$, and the result is the "B" form; but when the fiber is dry the hydrophobic forces are weaker, so the bread-and-cheese effect becomes the stronger of the two. Then the stacking involves a slide which is either positive—in which case the "A" form appears; or negative—in which case the "C" form appears. A left-handed "Z" form, mentioned briefly in Chapter 2, can also be seen in fibers when the conditions are relatively dry; and in this "Z" helix the base-pairs lie offset from one another as expected from theory.

Finally, according to Hunter and Sanders, there exists a third important kind of electrical force between base-pairs, which depends on the identities and locations of individual atoms in a base ring. Certain atoms, mainly hydrogens that are attached to nitrogens as NH or NH_2, or to carbon as CH, can generate a partial positive charge in their close vicinity; while other atoms such as oxygen O or simple nitrogen N can generate a partial negative charge. Thus, the stacking of different base-pairs on one another can be "fine-tuned" by the electrical interactions of individual atoms in the rings; and this feature will depend strongly on the sequence or ordering of bases at any step.

It follows from this that bases are not so free to slide over each other as we might suppose, on the basis of a model in which the surfaces of the bases are featureless, as we have assumed so far with our "block models." Instead, certain positions will be preferred—positions in which a positively charged atom from one base lies directly above or below a negatively charged atom from the other, and in which atoms with "like" charges do not lie directly above or below one another. In other words, the preferred positions will maximize attractive juxtapositions of atoms and minimize repulsive ones. This so-called "partial-charge" effect is well-known in chemistry; and it can have a striking influence in the positioning of large, flat molecules relative to one another.

Some accurately calculated partial charges for atoms in the DNA base-pairs are shown schematically in Fig. A3. These were provided by Hunter as a result of his computer calculations. Roughly, they correspond to what you would expect from having oxygens and nitrogens negative, and hydrogens positive; but the computer calculations are much more accurate than any assignment of partial charges that could be made from chemical intuition. For clarity, no individual atom types are identified;

but one may compare with Fig. 2.11(a) and (b) to identify the different kinds of atom at various places in the rings.

The most striking aspect of this figure would seem to be a dense accumulation of negative charge along the uppermost edge of the guanine ring, as shown at bottom left. Two negatively charged atoms, a nitrogen and an oxygen, lie there close together in space, and so generate a joint concentration of negative charge that is greater than that of any single atom elsewhere in the G–C or A–T base-pairs. What are the implications of a highly charged guanine ring for the three-dimensional structure of DNA? First, when two guanine rings stack over one another in a step GG/CC, they will not find it easy to stack vertically, one over the other at slide $S = 0$, owing to a strong repulsion between negatively charged regions; instead they will prefer to lie offset from one another at high slide, $S = +1$ to $+2$ Å (or

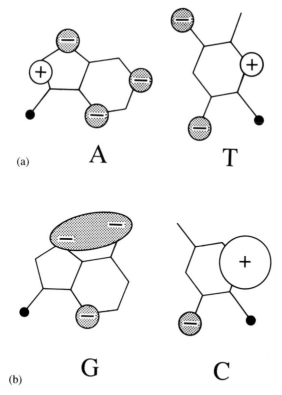

(a)   A   T

(b)   G   C

Figure A3 Regions of "partial charge" for AT and GC base-pairs. The base-pairs have the same relative orientations as in Fig. 2.11(a) and (b), where atom types H, C, N, O can be identified. Drawn from data supplied by Chris Hunter.

else at low slide, $S = -1$ Å), in order to minimize charge–charge repulsions. Second, when a guanine ring stacks over another base such as cytosine, as in the steps GC/GC or CG/CG, it may favor certain stacking geometries over others, because the strong negative charge of guanine will not fit equally well into all possible arrangements.

These theoretical notions are in good accord with experiment. Many DNA molecules containing GG/CC steps have been studied in the crystal or the fiber by X-ray diffraction, and wherever two or more GG/CC steps have been linked in series, as in the sequences GGG/CCC or GGGG/CCCC, almost always a high-slide structure has been observed. Also, the preferred stacking arrangements for GC/GC and CG/CG steps, as calculated by Hunter using the partial charges shown in Fig. A3, agree well with the stacking geometries for these steps as seen in DNA crystals.

In another case, the Hunter–Sanders theory predicts that the high propeller twist of A–T base-pairs, at sequences of the kind AA/TT, is due to a close contact between the thymine methyl group and a neighboring sugar. We said above that such high propeller twist was due to an additional hydrogen bond in the major groove, between one A–T base-pair and the next (see Fig. 3.6). In fact, the available experimental data do not distinguish clearly between these hypotheses; and we shall have to do additional experiments to learn which is correct.

In summary, the stacking interactions of base-pairs in DNA are influenced by three kinds of electrical force: (1) maximizing base-to-base overlap to avoid contact of the bases with water; (2) reducing base-to-base overlap to avoid repulsion of negatively charged surfaces; and (3) maximizing attraction and minimizing repulsion between partial charges on individual atoms in the base-pair rings. All of these effects will be subject, of course, to various constraints imposed by the sugar–phosphate chains of DNA; and so the base-pairs will not always be able to attain their optimal configurations relative to one another.

Nevertheless, we think that a deeper understanding of such forces will be of considerable use in understanding the actions of DNA in biology. For example, by altering the preferred values of roll, slide, and twist on a local scale, these forces can influence the outward appearance of the DNA in a way that may be recognized by a protein. Also, such forces are surely responsible for the influence of base sequence on DNA curvature, and for the easy unwinding of a TATA sequence as mentioned in

Chapter 4. But a complete discussion of these points is beyond the current scope of our book.

Further Reading

Hunter, C.A. and Sanders, J.K.M. (1990) The nature of pi–pi interactions. *Journal of the American Chemical Society* **112**, 5525–34. The three-part theory of Hunter and Sanders as applied to porphyrins.

Hunter, C.A. (1992) Sequence-dependent DNA structure: the role of base stacking interactions. *Journal of Molecular Biology* (submitted).

Answers to Selected Exercises

Note: most numerical values are given to more significant figures than are warranted by the data, in order to provide a better check on the arithmetic.

1.1 a Length of total DNA/diameter of cell = 200 000.
   b Volume of total DNA/volume of cell = 0.01.
   c Diameter of typical compact DNA ball = 6400 Å = 0.64 μm.
     Length of typical metaphase chromosome = 4.3 μm.

1.2 a Ala, Lys, Gln, Leu, Ile, Gln, Gly.
   b Pro, Ser, Asn, Ser, Phe, Lys.
     Gln, Ala, Thr, His, Ser, Arg.

1.3 a Ala, Lys, Gln, Arg, His, Ser, Arg.
   b Ala, Lys, Gln, Ser, Phe, Lys.

1.4 a Met, Ser, His, Gly, Thr, (Stop).
   b Met, Val, Ile, Arg, Asn, Ser, (Stop).

2.1 a 29°, 12.4 bp/turn.
   b 9.0 bp/turn.
   c 9.9 bp/turn.

2.2 a 330 Å, 10.0 turns.
   b 363 Å, 9.3 turns.
   c 495 Å, 6.4 turns.

2.5 a 2.
   b 21, 3.

3.1 Right-handed or clockwise.

3.6 "A":   4.7 Å,   22°, 2.5 Å
    "B":   0 Å,   0°, 3.3 Å
    "C": −2.2 Å, −9°, 3.1 Å.

4.2 b There are 3 AT-rich and 2 GC-rich regions.
   c There are 8 pyrimidine-purine steps.
   d GGCCC is the strongest region, and TATATA the weakest.

4.3 All bases except GCGC and CCGG at the two ends can make a cruciform about the central loop CTAG.

4.4 a 18, 10.8 m, 1.72 m.
   b 0.58 rad/m, 1.72 m.

4.5 (a) 30°, 0°, (b) 15°, −15°, (c) 9.3°, 0°, (d) 6°, −6°.

4.6 a 26.2°, **b** −26.2°, **c** 26.2°, **d** 30°, **e** 30°.

4.7 To adenine.

5.1 a −45°, 14 bp, −90 bp, 127 bp.
   b   45°, 14 bp,   90 bp, 127 bp.
   c   6°, 28 bp,   18 bp, 179 bp.
   d   84°,   3 bp, 178 bp, 179 bp.
Helix **d** has the smallest diameter, and helix **c** the largest. Helix **a** is left-handed, while helices **b, c, d** are right-handed.

5.2 a "A": 30.7°,   +11.2°
       "B": 36.0°,   0.0°
       "C": 39.4°,   −7.0°.
   b 34.0°.

5.3 b first sum = 24.9°, second sum = 8.1°, k = 26.2°, phase = 72°, maximum roll R is at step 2.

5.4 a 0.86 °/bp, step 6.
   b −1.7 °/bp.
   c 45 Å, 189 bp.

5.5 a 1.07 °/bp, step 7.
   b −1.5 °/bp.
   c 60 Å, 195 bp.

5.6 a 1.05°/bp, step 7.5.
   b 1.7 °/bp.
   c 50 Å, 180 bp.
A_4N_6 and A_6N_4 are left-handed, while A_6N_5 is right-handed.

6.1 a Once, approximately.
   b Once, approximately, depending on the kind of cord used.

6.4 a 2 turns, left-handed.
   b +1 turn, so that +1−2 = −1. To reduce the supercoiling or tangling of DNA on cell division.

6.5 a Tw = 0, −0.50, −0.87, −1.00 turns.
  Wr = −1.00, −0.50, −0.13, 0 turns.
  b −30°, −14°.

7.1 a 214 ± 13, 429 ± 15, 644 ± 20, 855 ± 21.
  b 214.5 base pairs.

7.2 a DNA occupies 53% of total volume.
  b 76 bp/hoop.

7.3 a **a** 291 Å. **b** 87 Å. **c** 100 Å.
  b **a** 300 Å. **b** 100 Å. **c** 50 Å.

7.4 a No.
  b No, they would join readily.

7.5 Nucleosome, 300 Å fiber, loops, metaphase coiling.

8.1 a Phosphorus, oxygen, nitrogen, carbon, hydrogen.
  b The phosphate PO_4 at lower-center left.
  c Because hydrogen scatters X-rays so weakly (36 times less strongly than carbon).
  d The phosphates.

8.2 a 5. **b** 6. **c** 10.

8.3 a Four kinds of fragment, of size 100, 200, 250, 450 bp.
  b 100 will be the fastest, and 450 the slowest.

8.4 a 250, **b** 1.5×10^7, i.e. 15 million.

8.5 Ratio of circumscribing volume to true volume = 3.2, for fiber radius 50 Å.

8.6 Wr = 0 for slowest; and since Tw = −12, then Lk = −12 for slowest (check: this is within the given range). Fastest has the value of Wr which differs most from zero; and since Tw = −12, the range of Wr is from +12 to −8 for Lk = 0 to −20. Therefore Wr = +12, Lk = 0 is the fastest.

Index